Ptilochronology

Oxford Ornithology Series
Edited by T. R. Birkhead FRS

1. *Bird Population Studies: Relevance to Conservation and Management* (1991)
   Edited by C. M. Perrins, J.-D. Lebreton, and G. J. M. Hirons
2. *Bird-Parasite Interactions: Ecology, Evolution, and Behaviour* (1991)
   Edited by J. E. Loye and M. Zuk
3. *Bird Migration: A General Survey* (1993)
   Peter Berthold
4. *The Snow Geese of La Pérouse Bay: Natural Selection in the Wild* (1995)
   Fred Cooke, Robert F. Rockwell, and David B. Lank
5. *The Zebra Finch: A Synthesis of Field and Laboratory Studies* (1996)
   Richard A. Zann
6. *Partnerships in Birds: The Study of Monogamy* (1996)
   Edited by Jeffery M. Black
7. *The Oystercatcher: From Individuals to Populations* (1996)
   Edited by John D. Goss-Custard
8. *Avian Growth and Development: Evolution within the Altricial-Precocial Spectrum* (1997)
   Edited by J. M. Starck and R. E. Ricklefs
9. *Parasitic Birds and Their Hosts* (1998)
   Edited by Stephen I. Rothstein and Scott K. Robinson
10. *The Evolution of Avian Breeding Systems* (1999)
    J. David Ligon
11. *Harriers of the World: Their Behaviour and Ecology* (2000)
    Robert E. Simmons
12. *Bird Migration* 2e (2001)
    Peter Berthold
13. *Avian Incubation: Behaviour, Environment, and Evolution* (2002)
    Edited by D. C. Deeming
14. *Avian Flight* (2005)
    John J. Videler
15. *Ptilochronology: Feather Time and the Biology of Birds* (2006)
    Thomas C. Grubb Jr.

Ptilochronology

Feather Time and the Biology of Birds

THOMAS C. GRUBB, JR.
Department of Evolution, Ecology, and Organismal Biology, Ohio State University, USA

and

Acopian Center for Conservation Learning, Hawk Mountain Sanctuary, USA

OXFORD
UNIVERSITY PRESS

OXFORD
UNIVERSITY PRESS

Great Clarendon Street, Oxford OX2 6DP

Oxford University Press is a department of the University of Oxford.
It furthers the University's objective of excellence in research, scholarship,
and education by publishing worldwide in

Oxford New York

Auckland Cape Town Dar es Salaam Hong Kong Karachi
Kuala Lumpur Madrid Melbourne Mexico City Nairobi
New Delhi Shanghai Taipei Toronto

With offices in

Argentina Austria Brazil Chile Czech Republic France Greece
Guatemala Hungary Italy Japan Poland Portugal Singapore
South Korea Switzerland Thailand Turkey Ukraine Vietnam

Oxford is a registered trade mark of Oxford University Press
in the UK and in certain other countries

Published in the United States
by Oxford University Press Inc., New York

First published 2006

British Library Cataloguing in Publication Data

Data available

Library of Congress Cataloging in Publication Data

Data available

Typeset by Newgen Imaging Systems (P) Ltd., Chennai, India
Printed in Great Britain
on acid-free paper by
Biddles Ltd, King's Lynn

ISBN 0–19–929550–6 978–0–19–929550–0

10 9 8 7 6 5 4 3 2 1

For Aidan, Finnegan, Fiona Claire, and Kelsey

Preface

This book is about the 18-year history of an idea. The idea came to me one February day in 1987 and has grown into what is now called ptilochronology, meaning, literally, the study of feather time. The name was inspired by dendrochronology, the study of tree time. The two "logies" are very similar in concept, but differ in scale; ptilochronology concerns daily intervals and dendrochronology yearly intervals. Since the publication in 1989 that introduced ptilochronology, the technique and its conceptual underpinnings have been applied in many parts of the world to a number of areas of avian behavior, ecology, and conservation biology. In this book, I review the diversity of concepts addressed by the method, consider how various criticisms have influenced development of ptilochronology, and suggest descriptive and experimental methodologies and analytical techniques to safeguard against invalid results. A final chapter sums up what this new technique has allowed us to understand about avian biology, and forecasts growing-points for future work.

In hopes of engaging the amateur student of birds, wherever possible I have tried to avoid technical jargon and the details of research design and analysis. However, such design considerations are sometimes required when validity of interpretation is in the balance. I have relegated scientific species names to an appendix.

Over the years since that day in February 1987, I have had the pleasure of discussing ptilochronology and doing science with many students and colleagues. Peter Stettenheim and Alan Brush were my mentors during the early days when I scrambled to understand how feathers grow. Fortunately, Alan, an expert on the morphology and physiology of feathers, was then the editor of *Auk*, the journal of the American Ornithologists' Union and one of the world's leading professional journals of ornithology. He recognized the promise of the initial article and gave it a home in his journal after it had been rejected elsewhere.

Mary Murphy and James King were tough early critics of the method and its assumptions. Their incisive remarks sharpened thinking about the limitations of ptilochronology and led to substantial improvements.

Arthur Wiseman's shoe-boxfuls of northern cardinal feathers were an early gold mine of information. Tom Waite and Olav Hogstad took a pioneer interest in experimental applications. Alan Carlson, André Dhondt, Jan Ekman, Olav Hogstad, Hans Källander, and Eric Matthysen hosted my visits to their universities, fed and sheltered me in their homes, and probed for weaknesses in the method.

Luc Lens, Ken Otter, and Toon Spanhove made available to me their unpublished datasets and results. In a spirit of open inquiry, Michael Kern invited me to examine on his flycatcher feathers the atypical growth bars that are still a complete mystery. David Cimprich, Paul Doherty, Andrew Dolby, Bob Mauck, Elena Pravosudova, Tom Waite, Reuven Yosef, and Greg Zuberbier all contributed to the development of ptilochronology in our lab while incorporating study of induced feather growth into their theses and dissertations.

Bruce Leach and Susan Ward at The Ohio State University and Janet Hinshaw at the University of Michigan located obscure library references and delivered crucial email attachments. Many of my colleagues have enriched this presentation with their photographs of feathers, birds, study sites, and researchers. Ian Sherman, Stefanie Gehrig, and Anita Petrie shepherded the manuscript through Oxford University Press.

Hiroyoshi Higuchi, Professor of Biodiversity Science, and his colleagues and students in the Department of Ecosystem Studies, Graduate School of Agriculture and Life Sciences, University of Tokyo provided an academic home for both stimulating conversation and the quiet intervals of scholarship during which most of the book was formed. I thank Stefan and Cathy Hotes and Ann and Kelsey Mongoven for their companionship in Tokyo. Ralph Boerner and Joan Herbers, my department chair and college dean, respectively, facilitated my stay in Japan. Through the years, I have benefited from many hours of discussion with my colleagues, Bob Mauck and Joe Williams. Finally, I thank Jill for her steadfast support.

Thomas C. Grubb, Jr.
November 2005

Contents

Part I Basic Ptilochronology

In this first part, I examine findings relevant to the interpretation of feather growth rates, the core concept of ptilochronology. Chapter 1 places ptilochronology in an evolutionary context, explains how the method works, and considers criticisms of some of its assumptions. Details of ptilochronology measurements and methods are presented in an appendix. Chapter 2 addresses the crucial question of whether reduced nutritional condition really does cause narrowed growth bars. Chapter 3 explores several possible alternative factors besides nutritional condition that could influence feather growth rates and thereby confound interpretation of feather growth records.

The second part of the book, termed Applied Ptilochronology, presents results of work using the method to investigate aspects of avian biology. While some of these projects concern bird responses to human-induced environmental change such as chemical pollution and habitat fragmentation, others explore ways in which nutritional condition, as indexed by rate of feather growth, can be used as a measure of fitness.

1 Introduction

1.1 *Ptilochronology, nutritional condition, and fitness*

One day in February 1987, I had an idea that became ptilochronology, literally the study of feather time. The essence of ptilochronology is a new method for determining the nutritional condition of birds free-ranging in their natural habitat. During subsequent years, dozens of workers have found this new measure useful in characterizing the responses of birds to various environmental factors, both harmful and beneficial. Now, some 18 years after the idea struck, seems a good time to take stock of what ptilochronology has taught us about the behavior, ecology, and conservation biology of animals.

I begin by making short digressions into the scientific method and evolutionary biology. With those items in place, I introduce the methods used in ptilochronology and what those methods produce. I also discuss ptilochronology in relation to other indices of nutritional condition. Chapters 2 and 3 consider various studies probing the utility and weaknesses of the method.

Before saying any more about ptilochronology, I would like to "prepare the ground" by reviewing basic tenets of the scientific method. The scientific method contains several components and processes. The first major component is termed the hypothesis, or model, and is a generality about the natural world that is too all-encompassing to be completely tested itself. The second component, the prediction, is deduced from the hypothesis and is sufficiently small in scope to be tested in the real world. The final component is called data, records, or observations. This component consists of facts resulting from simple observation of nature or from experimental tests of predictions deduced from hypotheses.

The process linking a hypothesis to a prediction is called deductive reasoning. There are rules for deductive reasoning which, if followed correctly, guarantee that if the prediction is shown to be false, the parent hypothesis must be false. If the prediction is shown to be true, the parent hypothesis may be either true or false. By finding the prediction true, we can only say that we have failed to disprove the hypothesis. The process for finding out if a prediction is true or false is called a test. Extensive guidelines exist for conducting tests, guidelines for assignment of replicates to treatment and control groups, appropriate statistical procedures, power analyses, and so on.

The process linking data to a hypothesis is called inductive reasoning. This is the process of combining various bits of information, various facts, in a new way to form a new generality. It is raw creativity. Hypothesis formation in science, the idea for the structure of DNA for example, is analogous to creation of the 12-tone scale in music, or impressionism in painting, or the Fosbury Flop in the high jump. Contrary to deductive reasoning and testing, there are

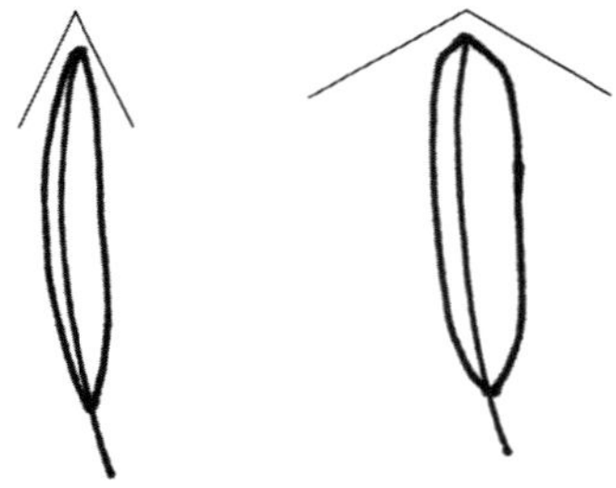

Figure 1.1 Tail feathers of typical juvenile and adult songbirds. The angle at the tip of the juvenile feather is much more acute than that of the adult feather. Such angles are used to age birds to juvenile or adult status. Tufted titmouse feathers collected to measure these angles gave rise to ptilochronology.

no rules for inductive reasoning. The only guideline seems to be that facts of various sorts need to be considered together. The party game where words representing disparate parts of speech are combined randomly to produce funny or instructive new sentences is a good example.

I have pursued this digression about the scientific method because first-person accounts reporting the genesis of a new hypothesis are not common. As I distinctly remember the way ptilochronology came about by inductive reasoning, I hope that relating the sequence of thoughts will be of some interest.

I had recently read that it was possible to age songbirds to either young-of-the-year or adult status by examining the angle made by the two parts of the vane at the outer end of a tail feather. This angle is more acute in young birds than in adults (Figure 1.1). In trying out this idea on the tufted titmice I was then catching and banding, I decided that I would pull out and save for study a standard tail feather, the outermost right, from each bird. (It was only later that I discovered it is a federal offense to possess feathers of most wild birds without state and federal permits!) I put each feather in a separate glassine envelope marked with the bird's band number.

Examining these feathers that evening under my desk lamp, I noticed a rather faint series of regularly alternating darker and lighter bands oriented almost, but not quite, perpendicular to the shaft, or rachis, of the feather. Under some forms of incident light, this banding or barring seemed to disappear altogether, and the pattern became most noticeable when a narrow beam of light struck the feather at an oblique angle and was reflected upward into my eye.

I had never noticed such barring before and set out to find out what I could about it. Thanks to key-worded electronic searches, it was not long before I discovered an article published in 1938 that demonstrated, by comparing repeated measurements of a growing feather on the bird with measurements of the width of individual bars on plucked feathers, that one such bar, termed a growth bar, denoted 24-hours' growth of the feather. This was fact number one.

It is fairly common knowledge that a bird will regenerate a feather to replace one lost, for example, during a territorial fight with a neighbor or a narrow escape from a predator. This was fact number two.

I was pruning Scotch pines on my farm one afternoon in February 1987, when, in a procession of mental thunderclaps, I was hit with the following thoughts:

1. Natural selection should favor the regeneration of lost feathers as rapidly as possible.
2. Feather regeneration requires energy and nutrients.
3. If such energy and nutrients are in short supply, the feather will, perforce, grow more slowly.
4. The growth bars tell me how much a feather grows during each 24-hour period.
5. By measuring the width of growth bars, I have a way of indexing how limiting were the energy and nutrients available for feather growth.
6. Because a bird carries a regenerated feather around with it until the next molt, it carries a multiple-day "printout" of its nutritional condition for me to collect whenever I next catch the bird.
7. If I first pull out an original feather and then pull out the regenerated replacement, I know the time interval in which the replacement was grown.
8. I can obtain an index of the nutritional effects of any particular environmental condition I measure while the feather was being grown. What follows is an elaboration of this series of ideas, this pulse of inductive reasoning.

By biological fitness, or just fitness, evolutionary biologists refer to the total lifetime reproductive success of a plant or animal. Fitness is central to all evolutionary biology. Actually, what we usually want to know about is called relative fitness. Organisms that are relatively more fit than other members of the same species pass more genes, and the characteristics producing those genes, to the next generation. For some reason, such individuals have been favored by natural selection, and their relatively greater transmission of genes causes the gene pool of the next generation to change, the definition of evolution. In theory, it is straightforward to measure relative fitness, you just count all the genes a plant or animal passes to the next generation and those that pass more genes are relatively more fit. (Continually mentioning both plants and animals is a bit cumbersome, so from now on I will let animals stand for plants as well and, pretty soon, I will dispense with all animals but birds.)

In practice, measuring relative fitness presents us with a whole rat's nest of problems. We cannot actually count all the genes or, properly, alleles of genes an animal contributes to the next generation in its offspring. Furthermore, animals share gene alleles with their non-offspring relatives by common descent, so if an animal's cousin, say, passes shared alleles to the next generation, the first animal is benefited. We need to sum the alleles an animal passes directly plus all of its alleles its relatives by common descent pass in order to calculate what William Hamilton (1964) termed its inclusive fitness. Then, we determine relative inclusive fitness by comparing our focal animal's inclusive fitness with the inclusive fitness of other members of its species.

But things are more complicated yet. What happens if an animal's offspring do not produce offspring themselves? That animal's long-term relative inclusive fitness might be very low (not zero if any non-offspring relatives produce offspring). Such complications as these have led a number of biologists to consider the number of grandchildren an animal produces as a good compromise indicator of its relative fitness. Number of grandchildren tells us something about both the survival and reproductive success of our study animals' offspring. Many field workers are very happy to get as far as being able to count relative numbers of grandchildren, let alone the grandnephews, grandnieces, cousins twice removed, and so on that would be required to really calculate relative inclusive fitness.

In most field studies, it is impossible to determine even the number of grandchildren, so workers have employed various fitness components, quantities that can be measured and that are assumed to be positively related to fitness. The growth rate of induced feathers, as determined by ptilochronology, is assumed to be a fitness component. A fundamental assumption behind the logic of ptilochronology is that there is a positive relation between the nutritional condition of a bird, as indexed by the growth rate of its feathers, and its biological fitness.

In employing feather growth rates, ptilochronology provides a sensitive indicator of subtle variation in fitness, variation not otherwise detectable. Let me try an analogy. Suppose we were nosy neighbors curious about the relative financial status of two households on our block. If one family lived in a peeling house with gutters falling off and had bad teeth and run-down shoes, we could be pretty sure it was in poorer shape financially. But suppose none of these indicators was revealing. We would need some more subtle indicator, something like the depth of tread of tires before they were traded in for new ones, or whether the kids went to private or public schools. These measures are not as readily apparent and, at the same time, indicate a smaller difference in economic well-being.

The assumption of ptilochronology about fitness is analogous to the assumption we make about the relation of tire tread depth to a family's financial fitness. For one reason or another, birds that are in better nutritional condition are likely to produce more genes (or offspring or grandchildren) than are birds in worse nutritional condition. This line of thought will shortly bring us to the definition of nutritional condition and alternative means of measuring it. As we will see, ptilochronology has certain advantages over other measures as a sensitive and valid indicator. Before comparing ptilochronology to other measures of nutritional condition, I introduce the logic and key elements of this recently invented technique (Grubb 1989).

Birds molt their plumage at yearly or more frequent intervals. However, when birds lose feathers by accident, such as during a near miss by a cat, replacements are promptly grown. Natural selection apparently favors expenditure of energy and nutrients to maintain a complete complement of body and flight feathers. Nevertheless, we know that feather growth can be reduced to compensate for nutritional demands of other activities. Molting periods tend to be separated in time from both breeding and migration. In some species, molting actually brackets the breeding or migration period. That is, feather replacement is retarded or suspended during times when feather production would compete for limited resources with other components of the annual cycle.

For several weeks after a feather is lost accidentally (or plucked by an experimenter), a new feather grows from the same follicle if the intake of energy and nutrients is adequate for feather formation. As such a feather grows from the collar zone of the follicle's papilla, growth bars are produced that are oriented at roughly a 90° angle with the feather's rachis. Each growth bar consists of one darker-appearing and one adjacent lighter-appearing band (Figure 1.2). Darker bands are apparently derived from material laid down in the follicle during the day and lighter bands consist of material laid down during the night (Wood 1950). Wood suggested that the difference in appearance of these bands is due to differences in optical properties brought about by differential incorporation of pigment into the keratin matrix of the feather during the day and night. This idea appears to be at least partially correct for the feathers of northern cardinals (see Appendix 2 for scientific names). By holding a northern cardinal feather in front of a light, it is easy to see the different amounts of pigment in darker and lighter bands. However, darker- and lighter-appearing bands are also found in white feathers such as the outermost tail feathers of dark-eyed juncos, and white feathers apparently do not contain pigments of any kind.

A very important assumption of ptilochronology is that, whatever the cause of the banding pattern, each pair of bands, that is each growth bar, constitutes a 24-hour period of feather growth. Michener and Michener (1938) provided initial evidence for this conclusion by measuring every day the growing tail feathers of house finches. After a feather was fully grown, they measured the width of daily growth bars and showed that growth bar width was the same as 24-hour growth measured as the feather was being grown. More recently, Anders Brodin (1993) has used radioactive labeling in an elegant confirmation of the assumption. He first plucked a tail feather from each of several willow tits. Several days later, after the induced feather had begun to grow out, he fed the birds several, and only several, sunflower seeds containing radioactive L-[^{35}S] cystine. Cystine is a sulfur-containing amino acid known to be incorporated into the keratin matrix of feathers. After putting the birds back on normal sunflower seeds, several days later he again fed them seeds with radiolabeled cystine. After the feather with the two radioactive cystine incorporations was fully grown, he pulled it and placed it on unexposed photographic

(a)
(b)
(c)

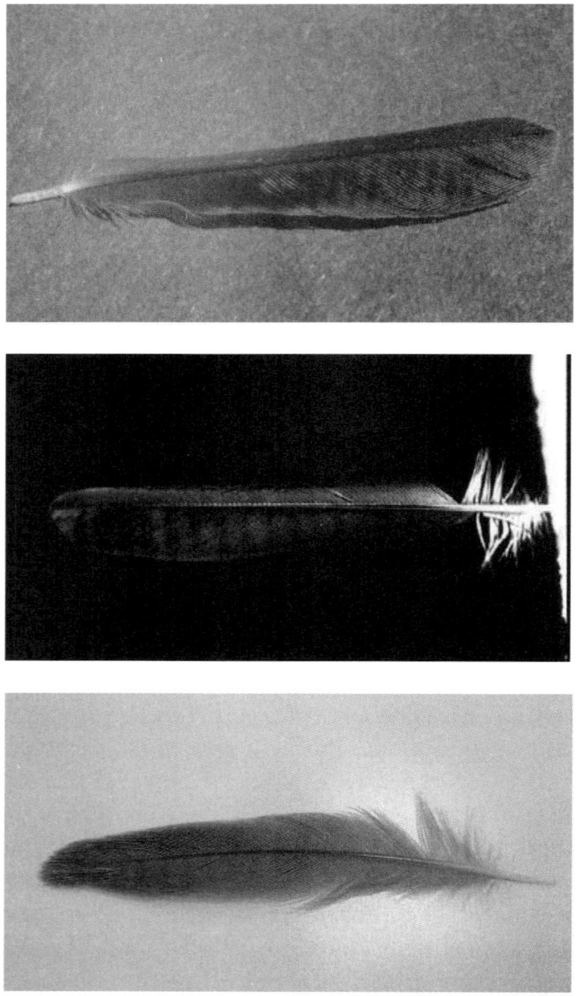

Figure 1.2 Tail feathers of (a) a sulfur-rumped flycatcher from Panama, (b) a great tit from Belgium, and (c) a pallid thrush from Japan. Each growth bar, consisting of one light-colored and one dark-colored band, denotes 24 hours of feather growth. Photograph (a) by José Diaz, (b) by Willem Talloen, and (c) by the author.

film. Radiation from the sites of ^{35}S incorporation left dark bands as tracings on the film (Figure 1.3.) Knowing the number of days between labeled-seed treatments and the distance between dark bands on the film, Brodin could predict the number of growth bars involved if one growth bar marked 24-hours of growth. The match between predicted and observed number of growth bars between the two radioactive bands was perfect. Production of growth bars seems a fundamental component of feather growth in a great many bird species and, with very few exceptions, it now seems safe to consider growth bars as indicators of 24-hour periods of growth.

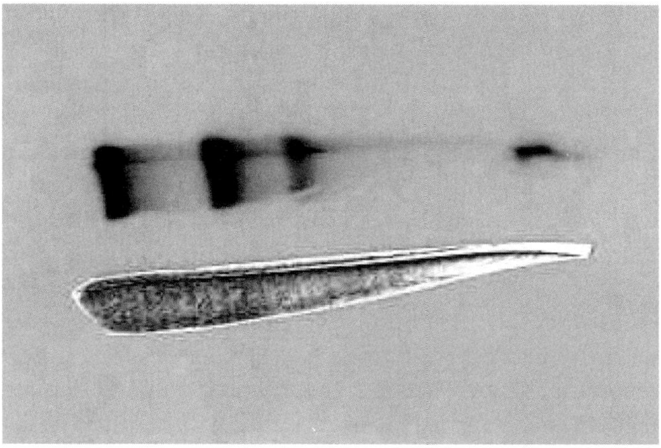

Figure 1.3 Induced willow tit tail feather photographed conventionally below, and by radiograph above. Dark bands on the radiograph denote the times when radioactive sulfur ingested in sunflower seeds was incorporated into the feather matrix. Photograph by Anders Brodin.

The breadth of a growth bar along the axis of a feather seems to be related to a bird's nutritional condition, and it is now time to discuss this sometimes-slippery concept (Grubb 1995). Definitions of nutritional condition are rare. Most workers appear to have avoided defining the term. Instead, they have considered the state of one or more components of an animal's body to be indices of nutritional condition (Harder 2005). However, quantifying an index is not a satisfactory substitute for defining what is being indexed. For example, in discussing the pitfalls of conventional indices of nutritional condition, King and Murphy (1984) pointed out that reductions in the body mass of wild animals have typically been attributed to a loss of "condition," but they did not define condition. In one of the uncommon attempts to come to grips with the term, Owen and Cook (1977) defined nutritional condition as "the fitness of a bird to cope with its present and future needs." They then concluded that variation in winter condition must be related to subsequent body mass, survival, and breeding performance. Thus, Owen and Clark had in mind that nutritional condition could be used to forecast lifetime reproductive success, or fitness. Following their line of thought, I define nutrition as the rate of ingestion of energy and nutrients that the body can assimilate. I then define nutritional condition as the state of body components controlled by nutrition and which, in turn, influences an animal's fitness. I consider nutritional condition to be synonymous with condition, body condition, and nutritional status. Nutritional stress results from an inadequate state of nutrition-controlled body components brought on by one or more nutritional constraints such as inadequate food supply and/or commitment of energy and nutrients to reproduction.

If you pluck a tail feather and release the bird, then re-catch the same individual more than a month later and pluck the newly grown replacement feather, the replacement (or induced) feather can provide a day-by-day record of the nutritional regime of the bird during the previous weeks. Individual growth bars (one dark band plus one light band) can be counted to establish the number of days taken to grow the feather. Matching induced feathers from two birds that had had feathers pulled at the same time and were then released in the same habitat could be used to determine the relationships of species, sex, age, social dominance status, kinship, and other factors to nutritional condition.

The induced feather is retained until the next molt. In most temperate- and boreal-zone birds, tail feathers are replaced annually during the so-called prebasic molt of late summer and autumn. A bird plucked in late autumn will grow the induced feather and will carry this record of nutritional status to be collected any time during the next 6–8 months. Appendix 1 is a detailed description of how ptilochronology measurements are obtained.

1.2 *Other indices of nutritional condition*

Other indices have been employed to measure nutritional condition as a fitness component. The more traditional and widespread of such indices monitor some aspect of body mass, such as total mass, fat-free mass, or percent fat. Other measures focus on the fat associated with certain organs (i.e. kidney; Riney 1955) or tissues (bone marrow; Hutchinson and Owen 1984). Components of the blood (Franzmann 1985) or urine (Del Guidice *et al.* 1989) have also been employed. Body mass and fat content have usually been viewed as indices of condition in the context of fecundity and starvation resistance. An animal that is more fecund or further from starvation has been considered to be in better nutritional condition, and as fecundity and resistance to starvation are often positively related to body mass, a heavier animal has been considered to be in better nutritional condition. For example, female waterfowl that were heavier upon arrival at the breeding grounds subsequently laid larger clutches (MacInnes *et al.* 1974).

The interpretation of correlations between these traditional morphological indices and fitness-related nutritional condition has been challenged (King and Murphy 1984). For example, it may not always be the case that animals with more body mass or fat content are more fit. Captive deer on *ad libitum* food lost body mass voluntarily at the onset of winter, a response interpreted as being adaptive (i.e. favored by natural selection) because it would reduce the energy required for maintenance metabolism over the winter (Warren *et al.* 1982). In some bird species, breeding females lose body mass in the absence of nutritional constraints at about the time their eggs hatch, apparently thereby reducing the energy cost of flight while commuting to the nest with food for the nestlings (Freed 1981; Croll *et al.* 1991). Such weight loss could also enhance the parent's fitness by permitting an increased proportion of food intake to be delivered to the young, resulting in heavier, better-surviving fledglings (Norberg 1981).

Body mass indices of nutritional condition can also be misleading because animals may reduce some body components while simultaneously augmenting others. During their flightless period, molting eared grebes at a migration stopover site significantly reduced the protein content of their large flight muscles while at the same time essentially doubling their subcutaneous fat (Gaunt *et al.* 1990). Here is a case where two commonly employed indices, fat-free body mass and percent body fat, would indicate opposite trends in nutritional condition. Shortly before departing southward later in the autumn, the grebes reversed the process by breaking down fat and building up flight muscles.

The theory of optimal body mass holds that for any animal there should be some body mass, neither too heavy nor too light, that would confer maximum fitness. From the perspective of such a theory, whole-body mass and fat content become ambiguous as indices of nutritional condition (Lima 1986; Rogers 1987). Increasing body mass (or fat content) is thought to decrease the risk of starvation, but at the same time increase the risk of predation because the greater mass may reduce maneuverability in response to an attack. Also, more time must be spent searching for and handling food at the expense of time spent watching for predators if a greater body mass (or fat content) is to be maintained. One prediction from optimal body mass theory is that animals with unpredictable food supplies should maintain larger fat reserves as insurance against starvation than should animals with relatively stable food supplies (Rogers 1987). Supporting this prediction, among songbirds wintering in the temperate zone, ground foragers carried more fat than tree foragers. Ground foragers, but not tree foragers, could have their food supply cut-off by snow cover at any time (Rogers 1987).

The view that animals may maximize their nutritional condition by maintaining an optimal body size or fat content suggests an important reason why nutritional indices besides feather growth rate could be ambiguous. Consider two members of the same species from the same habitat that are the same size structurally (same size skeleton), but that differ in body mass or fat content. The heavier or fatter animal could be in either better or worse nutritional condition than the lighter one. The heavier animal could be at maximal nutritional condition and the lighter animal at higher risk of starvation, or the lighter animal could be at maximal nutritional condition and the heavier animal could be compensating for lower predictability of its food supply. This ambiguity of interpretation is illustrated in Figure 1.4a and c.

For several reasons, components of blood and urine also appear to be dubious indices of nutritional condition. Blood components are subject to poorly understood daily cycles and are affected by stressors, some of which are related to sample collection. Harder (2005) points out other complexities. For example, urea nitrogen, an index of protein breakdown found in the blood and urine, has been found to decrease initially and then increase as an animal

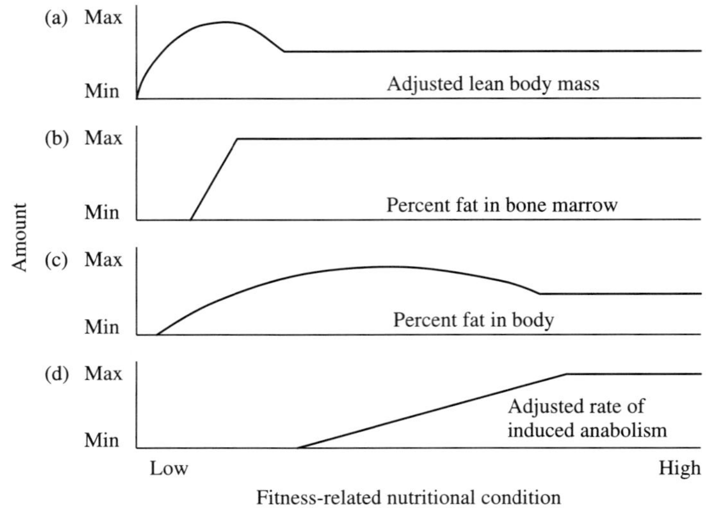

Figure 1.4 Presumed relationships of several indices to fitness-related nutritional condition. From Grubb (1995).

begins depleting its protein stores. If animals maintain an optimal fat-free body mass (Figure 1.4a), then we can imagine complex and ambiguous relationships between levels of blood or urea nitrogen and nutritional condition.

Aside from the induced feather growth to which I will turn shortly, there appears to be at least one other unambiguous index of nutritional condition, namely fat content of bone marrow. A low value of this index appears to be a very good predictor of death by starvation (Hutchinson and Owen 1984). However, the utility of this index is quite limited because an animal does not break down marrow lipids until it is near starvation, by which time reduction in nutritional condition has become substantial and obvious (Figure 1.4b). Stated another way, there appears to be no ambiguity of this measure because the optimal bone marrow content is the maximum, and the maximum is maintained until other reserves (i.e. non-marrow fat, protein) have been depleted.

1.3 *Feather growth as an index of nutritional condition*

Induced feather growth can be thought of as a variety of induced anabolism, a term describing the induced regeneration of body tissue by a plant or animal. The utility of induced feather growth as an unambiguous index of nutritional condition stems from asking the following question. After having allocated the required amount of energy for maintenance and the optimal amount of energy for reproduction, how quickly can you regenerate a feather? This question rests on the assumption that natural selection has favored the regeneration of lost body parts as rapidly as possible.

Induced feather growth appears to be an unambiguous index of nutritional condition (Figure 1.4d). Consider again two animals of different body mass. If they are birds of the same species and we control for effects of age, sex, and season on the rates at which they grow an induced feather (Grubb *et al.* 1991), we should be able to rank them according to their nutritional condition. The induced-feather-growth model (or, more generally, the induced-anabolism model) predicts that regardless of body mass or fat content, the animal in better nutritional condition should regenerate a feather at a faster rate.

For several reasons, the regeneration of a feather may be a type of induced anabolism uniquely suitable to manipulative experiments. First, a full-grown feather is held within the follicle solely by dead connective tissue, so inducing the regeneration apparently results in little or no discomfort to the animal. Second, the loss of one tail feather, the type usually plucked, probably has little effect on a bird's ability to fly, and has essentially no effect on the insulation or water-repellency of the feather coat. Thus, the removal of the feather has little impact on fitness independent of the energy and nutrients its re-growth demands. Third, even after the induced feather is fully grown, it can be divided into daily growth increments, so multiple captures are not required to determine rate of re-growth. Fourth, in the majority of bird species, an induced flight feather is retained until the next pre-basic (autumnal) molt, so a record of nutritional condition can be collected weeks or months after the original feather has been pulled.

While beyond the scope of this book, rate of regeneration of body parts can be extended conceptually to rate of replacement of lost products such as squid ink , skunk oil, or nectar, lost caches such as those made by squirrels or northern shrikes, or lost construction such as termite tunnels or beaver dams (Grubb 1995). So far as I know, inducing these other products, caches, or construction as a means to investigate nutritional condition and fitness has not been attempted.

1.4 *Criticisms of ptilochronology*

Soon after I introduced ptilochronology (Grubb 1989), Mary Murphy and James King (1991) published a substantial critique of the method, pointing out what they believed to be untested assumptions and limitations. In the sections to follow, I have often modified my initial response (Grubb 1991) to Murphy and King's critique in light of subsequent findings. Later papers by White *et al.* (1991) and Jenkins *et al.* (2001) also contain evaluations of ptilochronology's utility. Although we will return later to other material in the latter two articles, it seems appropriate to consider their evaluations of ptilochronology here together with Murphy and King's (1991) paper.

Although Murphy and King felt that their feeding trials with white-crowned sparrows were germane to ptilochronology, they calculated daily growth of flight feathers from periodic measurements of the growing feather on the bird rather than from the actual width of growth bars. They said they did not measure the width of growth bars because "they were difficult to see and often invisible on *Z. l. gambelii*." Apparently, this statement applies both to the original feather grown during the preceding molt and to the induced feather grown during the feeding experiments. Lack of discernable growth bars on the original feathers of their birds is puzzling. I found discernable and measurable growth bars on the outermost right tail feather of all 14 *Z. l. gambelii* in the collections of The Ohio State University Museum of Biological Diversity.

While growth bars are usually quite apparent on newly induced flight feathers of sparrow-sized birds, Murphy and King were not able to see them clearly enough to measure. Growth bars are difficult to see on abraded feathers and the tail feathers of birds housed in the very small cages they used were "often frayed or broken." In an experiment demonstrating that a deficient diet caused narrower growth bars in Carolina chickadees (Grubb 1991), an experiment to be discussed in detail later, I purposely used relatively large cages to reduce abrasion of induced feathers.

It may be important to note that Murphy and King actually caught their sparrows every three days to measure the growing feathers. In earlier work, they had demonstrated that handling sparrows provoked fault bars, areas where production of feather barbules is deficient, producing a "thin spot" on the feather (King and Murphy 1984). Whether handling birds often while they are growing feathers also affects size and distribution of growth bars is currently unknown. However, a strength of ptilochronology is that it does not involve handling a bird at all while it is growing a feather. A very nice project would involve comparing growth bar production on induced feathers as a function of handling frequency of the bird during feather growth, with a control group not handled at all during the period.

Recently, Daniel Cristol and his students have used captive white-throated sparrows, a species in the same genus as Murphy and King's birds, to provide strong evidence that an inadequate diet does cause a reduction in growth bar width (Jenkins *et al.* 2001). These workers randomly assigned 36 male white-throated sparrows to each of two groups, housing each group communally in a large aviary (Figure 1.5). Birds in one group received a subsistence diet, 200 g/day of corn and millet. The other group received substantially more and higher quality food, 500 g/day of corn, millet, sunflower-seed hearts, thistle seeds, high-protein turkey-starter and 2–3 mealworms per bird. In early February, outermost tail feathers were plucked from all birds in both groups and 52 days further along, induced feathers were taken.

All birds regenerated a replacement feather. While growth bar width and total length of original feathers did not vary statistically between low- and high-diet birds, high-diet birds had significantly longer induced rectrices marked by significantly wider growth bars (Figure 1.6). Thus, in this study of a close relative of Murphy and King's birds, growth bars were readily seen and growth bar width was clearly related to adequacy of diet.

(a)

(b)

Figure 1.5 (a) White-throated sparrow and (b) aviary used in feeding trials of captive flocks of this species. Photographs by Daniel A. Cristol.

Murphy and King (1991) noted a number of aspects of ptilochronology that they considered to be untested assumptions. The first assumption they listed was that each growth bar actually marks 24 hours of growth. At the time, we had only the early measurements of Michener and Michener (1938) and Wood (1950) as evidence for this assumption. With Brodin's (1993) radio-labeling study, detailed earlier, this assumption has now been substantially supported. Still, several studies to be considered further on appear to be exceptions to the generality.

The next assumption Murphy and King noted was that growth rate and final length of feathers were both predictable from the extent of malnutrition, regardless of the type of malnutrition. In my 1989 paper, I was careful to

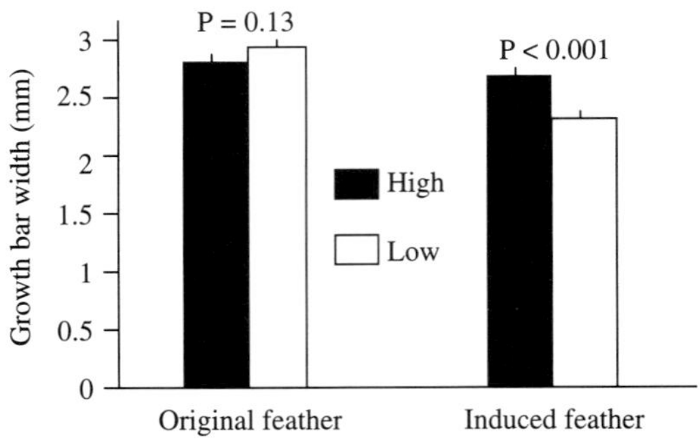

Figure 1.6 Original and induced growth bar widths (± SE) for captive white-throated sparrows fed on ample diet (High, n = 36) or a subsistence diet (Low, n = 35). Probabilities that observed differences could have occurred by chance are shown above the bars. From Jenkins *et al.* (2001).

maintain that reduced nutritional status was only a sufficient cause of reduced growth bar width. We now have considerable evidence that factors such as day length and associated hormonal changes and age and sex of a bird can also have an effect on feather growth. Also, in my introductory paper, I did not assume any effect of nutritional status on the total length of an induced feather. Apparently, there sometimes is such an effect (Grubb 1989; Waite 1990; Grubb *et al.* 1991) and sometimes not (Grubb and Cimprich 1990; White *et al.* 1991).

Murphy and King thought the method assumed that feather growth rate slows immediately with the beginning of any nutritional shortage. I did assume that the effect of such a shortage would likely be detected on a daily basis, that is, would occur within 24 hours. In practice, however, averaging the widths of a number of growth bars, 10 in the original method, reduces the necessity of assuming immediate reduction in growth with malnutrition. Nevertheless, it would be worthwhile to have carefully controlled information from the laboratory focused on time lags between reduced nutrition and reduced nutritional condition as indexed by feather-growth rate. We now have some evidence from a study of nestling tree swallows in nature, to be discussed later (McCarty and Winkler 1999), that time lags may occur between food input and feather growth rate.

Murphy and King held that implicit in the logic of ptilochronology was the assumption that feather growth slows in direct proportion to the magnitude of the nutritional shortfall. We now have two pieces of information suggesting there is some threshold degree of shortage that must occur before the rate of feather growth will be slowed. First, in their feeding trials with white-crowned sparrows, Murphy and King found that depleting a synthetic diet by half in certain amino acids failed to slow feather growth. Such results with synthetic diets are difficult to interpret because, in nature, any diet adequate in energy also is likely to be adequate in specific amino acids (King and

Murphy 1984). Second, in a controlled deprivation experiment with Carolina chickadees that I performed (Grubb 1991), there appeared to be some threshold degree of deprivation before feather growth began to diminish. Thus, Murphy and King's criticism seems to be correct; the implicit assumption of universal dose dependence of feather growth rate on level of nutrition does not hold. In experimental work, where induced feather growth is used to compare the nutritional condition of different treatment groups, it now seems possible that all treatment groups could be above the threshold value necessary for reduction in feather growth to occur.

Murphy and King made an important point in stating that an assumption implicit in ptilochronology is that daily growth rate of the original feather occurred "when the birds were well-nourished and, if two or more birds are being compared, in otherwise identical environmental conditions." Difference in growth bar width on feathers of two birds can potentially be confounded simply by differences in structural size. All else being equal, bigger birds are likely to grow feathers with wider growth bars. As noted earlier, daily growth of the original feather has been used to standardize growth rate of an induced feather grown from the same follicle. That method does assume that the original feather was grown during the normal molt. Feathers not grown during the molt can be detected by their unusual color, degree of abrasion, and/or growth bar width.

The original method did not assume that a bird was "well-nourished" during growth of an original feather, only that birds in different treatment groups were equally nourished. As it turns out, part of Murphy and King's cautionary statement was correct; we now know that growth rates of original feathers can vary among birds (see, for example, Chapter 7 on individual quality).

Murphy and King stated that ptilochronology assumed that periods of nutritional shortfall always coincided with the standard segment of growth bar measurement. This criticism seems quite similar to the one mentioned earlier about lack of lag time between nutrient shortfall and reduction in feather growth rate. In any case, the method of determining daily feather growth consists of finding the average width of a number of growth bars located in a standard position on a feather. The method assumes that a narrowing of growth bars within this feather segment indicates lowered nutritional condition, but would fail to detect episodes of reduced nutrition coinciding with growth of other parts of the feather.

The last assumption pointed out by Murphy and King was that any reduction in feather growth rate would indicate a reduction in nutritional condition sufficient to reduce fitness, or lifetime reproductive success. My intention was to suggest that, all else being equal, reduced growth bar width indicates the possibility of a nutritional challenge sufficient to affect a bird's fitness. That is, I accepted the possibility that some reduction of growth bar width could indicate a reduction in nutritional status too insignificant or transitory to cause a reduction in fitness. In rereading my introductory (1989) article, I realize now that I did not make this point as clearly as I could have.

The possibility that reduced growth bar width indicates reduced fitness is, of course, what many researchers have found heuristic about ptilochronology. Therefore, Murphy and King's conclusion that this assumption is "perhaps untestable" is a serious criticism. By "untestable," I assume that Murphy and King mean that the assumption cannot be proven false. Some reduction in growth bar width may occur under environmental conditions not severe enough to produce breakdown of body tissues (the conventional measure of nutritional stress; King and Murphy 1984), let alone to reduce fitness. However, if no reduction in some established fitness component, such as survivorship or lifetime reproductive success, occurred over many trials separated in space and/or time under conditions that substantially reduced growth bar width, then the ecological and evolutionary implications of the technique would be gravely weakened.

At the time it was published, the criticism was highly justified as any relationship between rate of feather growth and fitness was then quite speculative. However, as we will see in the chapters to come, we now have evidence linking rate of induced feather growth to both survivorship and reproductive success.

The critical review of Murphy and King was quite useful in advancing conceptual thinking about ptilochronology. Partially as a result of their comments, much of the research to be discussed has employed safeguards against confounding variables. For example, attempts have been made to obtain sample sizes sufficiently large to reduce the effect of any unidentified atypical original feather. In manipulative studies, treatment and control trials have been arranged in a balanced design, thus avoiding potentially confounding effects of seasonal variation and, in the best-designed studies, birds have been randomly assigned to treatment and control groups.

Murphy (1992) provided a series of final reactions in our published exchange, most of which were reiterations of the points King and she made in 1991. So far as I am aware, her call for research into the proximate causes of the formation of growth bars has gone unheeded.

White *et al.* (1991) used ptilochronology to investigate reproductive effort in breeding birds, a study that I will consider in detail later. Of interest here are their general concluding remarks on strengths and weaknesses of the method. In their experiment, two birds failed to grow an induced feather at all and several delayed the onset of induced growth. The researchers pointed out that the delays, in particular, could confound studies of short-term reductions in nutrition. This is a valid point to which might be added that any delay in onset of induced growth during a short-term deprivation trial of some kind might result in a feather with wider growth bars being grown after the short-term deprivation had passed. In an experiment, such a delay would tend to bias results against a difference between groups as a result of the treatment.

White *et al.* made the point that if success of recaptures required to collect an induced feather varies among treatment groups, a bias could be introduced. This potential confound is readily detectable by monitoring recaptures. Their final point was that since the relationship between daily growth of an induced feather and that feather's total length is uncertain, in cases where growth bars are indistinct, use of total feather length as a measure of nutritional condition may not be valid due to slower-growing feathers growing for more days. Again, such compensatory growth would tend to bias against detecting differences among treatment groups.

In concluding the discussion of their study of diet and feather growth, Jenkins *et al.* (2001) made a distinction between ptilochronology's abilities to detect density-dependent and density-independent ecological stressors. A density-dependent factor is one whose impact is correlated with density. Density is measured as items per unit area, such as birds per km^2. For example, the effect of a limited supply of food is density dependent. The higher the density of animals, the more intense the adverse effect on each animal. By contrast, a density-independent effect is one whose impact does not depend on animal density. For example, the killing effect of a cold rain on small birds when they are roosting at night is independent of the density of those birds. Jenkins *et al.* argued that while density-dependent factors such as territory size, social rank, and degree of ecological competition have been shown to effect feather growth, various density-independent factors, wind-chill and day length, for example, have not always done so. While the universality of this distinction is not yet known, their point is well taken that employers of ptilochronology should be aware of whether ecological factors they wish to examine can have any effect at all on feather-growth rates.

1.5 *Summary*

Chapter 1 began by placing ptilochronology and the growth rates of feathers within the framework of direct and inclusive relative fitness. The practical aspects of using feather growth rates to index nutritional condition were described, and ptilochronology was put forward as having several advantages over previously existing measures of nutritional condition. The chapter concluded with discussion of three articles containing critiques of ptilochronology. Certain of the points raised in those publications, such as the issue of time lags, the question of dose-dependency and dosage thresholds, differences in growth rates of original feathers, and whether feather growth rates are sensitive to only density-dependent environmental factors, have influenced subsequent study designs even when not well supported themselves. In the next chapter, we continue the study of basic ptilochronology by probing for the utility and ambiguities associated with the technique.

2 Can reduced nutritional condition really cause narrowed growth bars?

Ptilochronology rests on the assumption that rate of feather growth is directly related to a bird's nutritional condition and that, within limits, nutritional condition is related to net rate of energy intake. The more food, the better the nutritional condition and the greater the rate of feather growth. As Murphy and King (1991) stated in their evaluation of the method, it would be reassuring to have evidence supporting this assumption. I begin by providing experimental evidence with results from laboratory feeding trials of my own most frequently used study species, the Carolina chickadee (Grubb 1991, Figure 2.1a). I thank Robert A. Mauck for his help with this experiment. We tested the prediction that a known reduction in food intake below the *ad libitum* level would cause a reduction in the width of growth bars on induced feathers.

In January 1990, we captured 15 chickadees and introduced them singly and at random to cages arranged in a 3 × 5 grid within a windowless building (Figure 2.1b). All birds were housed at the same ambient temperature and under the same January regime of 8 hours of light and 16 hours of darkness. For several weeks, we determined each bird's individual *ad libitum* consumption of sunflower seeds and mealworms. We did this by subtracting the weight of food left in the food dish at the end of the day, after the chickadees had gone to roost for the night, from the weight of food at the beginning of the day, before the birds had begun to feed. Every day, we also carefully sorted through the material on the bottom of the cages, picking out bits of uneaten food to be subtracted from the amount of food removed from the food dish.

Once we had determined the food consumption of each bird under *ad libitum* conditions, we randomly assigned five of the 15 birds to each of three treatment groups, 100%, 90%, or 80%, by weight, of the individual bird's *ad libitum* food consumption. Then, we pulled and stored in a glassine envelope the outermost right tail feather of each bird.

Over the next 6 weeks, just before "lights on" each day, we gave each bird its assigned amounts of seeds and mealworms and a fresh supply of water. After the 6 weeks had passed, we pulled the induced feather, allowed the deprived birds several days of *ad libitum* food consumption, then released each bird at the site where we had captured it.

We measured average growth bar width as noted earlier and, to standardize for bird size, divided the average daily growth of the induced feather by the average daily growth of the original feather. We also obtained the mass of each feather to the nearest 0.01 mg on an electronic scale and divided induced mass by original mass to control for bird size.

(a)

(b)

Figure 2.1 (a) Carolina chickadee, and (b) Bob Mauck with cages holding visually isolated Carolina chickadees growing induced feathers while on a diet of 80%, 90%, or 100% of their *ad libitum* diet. Photographs by the author.

As predicted, there was a direct relation between proportion of *ad libitum* diet and width of induced growth bars as a proportion of the width of the original growth bars. A statistical test showed that this difference was significant between the 80% group and both the 90% and 100% groups, but not between the 90% and 100% groups (Figure 2.2). Similarly, the mass of the induced feather as a proportion of the mass of the original feather was significantly lighter in the 80% group than in the 90% group, while the 90% and 100% groups were about the same.

The results of this deprivation experiment furnish strong support for the assumption that feather growth rate is a valid index of nutritional condition. While the effect of reduced food on feather growth was only minor in the 90% treatment group, the effect was clear in the 80% group.

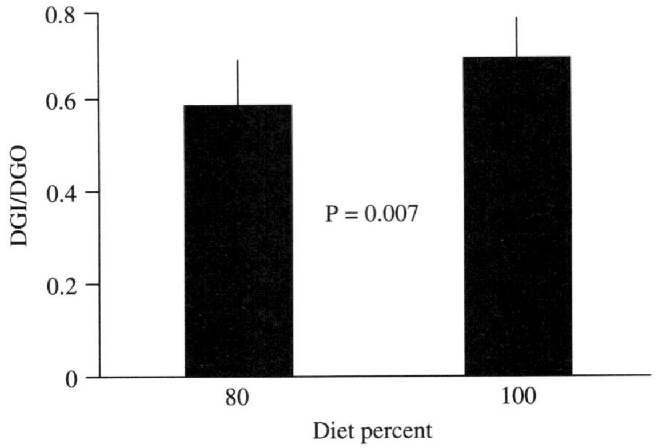

Figure 2.2 Mean (± SD) ratio of daily growth of induced (DGI) and original (DGO) R6 tail feathers of captive Carolina chickadees fed either 80% or 100% of their *ad libitum* diets. The probability that the observed difference could have occurred by chance is shown between the bars. From Grubb (1991).

A second message from this experiment is that other factors besides net energy and nutrient intake can play a role in determining the daily rate of feather growth. Birds in the 100% treatment group were maintained on an *ad libitum* diet and in their thermoneutral zone (the zone of minimum expenditure of metabolic energy) where, by definition, the energy requirement for existence was at a minimum. Yet these birds grew an induced tail feather at a daily rate only 73% as great as the daily rate they had grown the original feather during the previous molting period. We must conclude that besides present nutritional condition, daily feather growth in captivity is apparently responsive to other factors, some of which may vary through the year. We will further explore such circannual factors shortly.

Besides daily feather growth, induced feather mass as a proportion of the mass of the original feather varied significantly with percent of *ad libitum* food. This result suggests that the total quantity of material incorporated into a feather is also responsive to a bird's nutritional condition. As we shall see, some workers have used this ratio of feather masses to monitor nutritional condition in systems where growth bars on induced feathers are difficult to discern.

Recently, Daniel Cristol and his students (Jenkins *et al.* 2001) have essentially repeated our chickadee lab study with white-throated sparrows, also corroborating the positive relation between food adequacy and feather growth bar width. Using only male sparrows, they created flocks of 36 birds in each of two large outdoor aviaries. The "Low food group" was maintained each day on 200 g of a mixture of four seed types, while the "High food group" was maintained each day on 500 g of the seed mix plus four mealworms per bird and *ad libitum* access to carrots, boiled eggs, and parsley. In short, in the High food group, preferred foods were available to all birds at all times. In early February, the researchers pulled the two outermost right rectrices from all birds. Fifty-two days later, they pulled the induced feathers.

There was no difference between the Low and High groups in growth bar width of the original feathers. However, when Low and High groups were compared in a multivariate analysis using growth bar width on original feathers as a covariate, growth bar width of the High group was significantly greater (Figure 1.6). It should be noted that this result is weakened by the fact that the 36 birds in each treatment group were all housed together in the same cage so should not be considered to be statistically independent. A statistical extremist would hold that the sample size in High and Low groups was 1, the number of cages, not 36, the number of birds.

David Cimprich and I supported the manipulative lab studies with records from an observational study of birds in nature (Grubb and Cimprich 1990). Here, a balanced design with randomly assigned control and treatment groups was not possible, but it was comforting to have wild, free-ranging birds furnish similar results. We had been studying mixed-species flock associations among the four common wintering woodland birds in Ohio, Carolina chickadee, tufted titmouse, white-breasted nuthatch, and downy woodpecker. In five of the woodlots where we were working we left feeders in place all winter. In another five woodlots, we used feeders only as long as required to catch and band the birds, then left the feeders in place, but empty for the winter. We asked whether birds in the woodlots with maintained feeders would grow induced feathers more rapidly than those in unsupplemented woodlots.

In December, we trapped and banded birds in both types of woodlots, pulled the outermost right rectrix, then let the birds go. (The outermost rectrices of woodpeckers are actually small semi-vestigial remnants, so for the downy woodpecker, we pulled the second-most-outer right rectrix.) At the end of February, we refilled the empty feeders and collected the induced rectrix from birds recaptured at provisioned and unprovisioned sites. As in the chickadee lab experiment, we calculated the proportion of the growth bar width of original feathers represented by the growth bar width of induced feathers. We then compared these proportions between provisioned and unprovisioned birds of each of the four species, subdividing results by sex and age. (Some of these birds can be assigned as either juvenile, that is, born the previous summer, or adult, of undetermined age, but at least 1 year older than juveniles.) Records of all birds of the same age and/or sex in the same woodlot were averaged so, to use the statistical term, the woodlot was the primary sampling unit.

In all bird categories, daily growth of the induced feather during the winter was considerably slower than that of the original rectrix that had grown during the previous summer or autumn. The proportion of induced daily growth varied from 64% in unprovisioned immature chickadees to 80% in provisioned male and female nuthatches. In every one of the 10 age or sex categories where a comparison could be made, the proportion was larger for

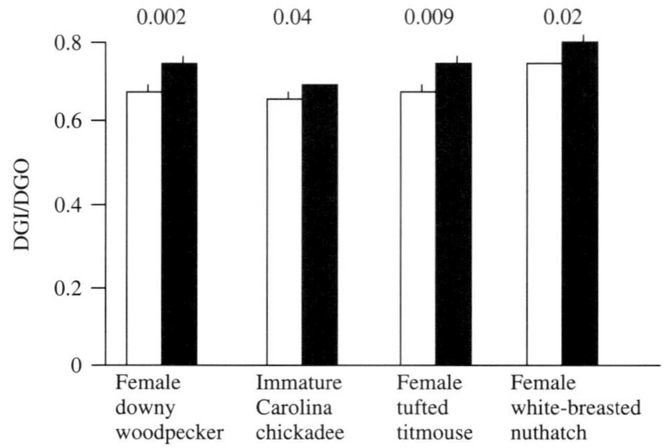

Figure 2.3 Ratio (± SD) of daily growth of induced (DGI) and original (DGO) R6 tail feathers of woodland birds wintering in habitats unsupplemented (open bars) or supplemented (filled bars) with *ad libitum* sunflower seeds. Probabilities that observed differences could have occurred by chance are shown above the bars. From Grubb and Cimprich (1990).

supplemented than for unsupplemented birds. In four cases, the extent of the difference reached statistical significance (Figure 2.3).

A few years after our woodlot study, Jan-Åke Nilsson found a similar effect of provisioning on feather growth of European nuthatches in his study sites near Lund, Sweden (Nilsson 1994). Nuthatches in some woodlots were given sunflower seeds for the month of November only, and those birds cached large numbers, perhaps hundreds, of the seeds in bark-crevices of surrounding trees. Nuthatches in other woodlots were left unprovisioned. Between mid-January and the first day of February, 18 birds from the November-provisioned woodlots and 16 from unprovisioned woodlots were captured. After their right and left fourth rectrices had been plucked, the birds were released. (In this nuthatch, the fifth and sixth rectrices are pied black and white, making the reading of growth bars difficult. The fourth rectrix is uniform in color and a good deal easier to read.) The nuthatches were recaptured in March and early April and their induced rectrices collected.

In agreement with previous studies, birds in the provisioned woodlots had longer induced feathers with wider growth bars. Another feature of this result is that the effect of the provisioning on feather growth was apparent several months after the provisioned food had been removed. Such a strong delayed response hints at the adaptive significance of caching as a life-history strategy, a matter we will return to in Part II.

Actually, the major thrust of Nilsson's 1994 paper was a test of fluctuating asymmetry of tail feather length as a function of nutritional condition. We will focus on fluctuating asymmetry a bit later on, but it is worth noting here that Nilsson used feather growth rates as his benchmark of nutritional condition against which to assess predictions about asymmetry. By 1994, workers were beginning to place confidence in the validity of the technique.

2.1 *Summary*

This chapter has focused on one observational and three manipulative projects, two in the lab and two in nature. All supported the assumption of ptilochronology, that rate of feather growth is positively related to food supply. Clearly, though, a proximate effect of nutritional condition is not the entire answer. If it were, induced feathers grown on *ad libitum* food at other times of year, winter in the case of these studies, would have growth bars as wide or wider than those of feathers grown during the normal molting period in late summer and early fall. That they do not suggests that other causal factors are in play. In the next chapter, we delve further into what these other factors might be.

3 *Does any factor other than nutritional condition control feather growth?*

3.1 *Annual cycles*

In Chapter 1, I argued that reduced nutritional condition could be a sufficient cause of reduced feather growth rates. I now consider several projects that searched for other factors that could cause variation in feather growth rate. Knowing the impact of such factors might help us decide how freely we may invoke poor nutrition as the cause of narrower growth bars. Three relevant studies are available, a descriptive account of circannual rhythms and two manipulative laboratory studies, one addressed to ambient temperature or wind chill, and the other to follicle history.

In early 1990, I presented a talk on ptilochronology at the Cincinnati Zoo. After my talk, Arthur Wiseman, a retired pharmacist, approached me saying he was an amateur bird-bander and that for a number of years he had been pulling the two outermost tail feathers from northern cardinals he captured. Furthermore, in many cases, he had pulled original feathers and then, weeks or months later, had pulled the induced feathers from the same follicles. Would I be interested in examining such feathers? I was dumbstruck by such good luck. Once the growth bars on Wiseman's feathers had been measured, we would have an extensive data set collected over most of the year with which to investigate questions about the annual cycle of induced feather growth.

Before I get to the annual-cycle results, you might be interested to know why Wiseman collected all those feathers. He was attempting to answer a simple and intriguing question. It is well known to ornithologists that a bird in the hand can often be aged to juvenile (i.e. young-of-the-year) or adult status by the angle that the terminal (i.e. distal) edges of the vanes of a tail feather make with the rachis, or shaft. That angle is more oblique in adults than in juveniles (Figure 1.1). The reason for the difference is not known, but may be related to the fact that growth of the first rectrix (started in the nest and finished after a juvenile bird fledges) permanently expands the follicle so that the tips of succeeding feathers from the same follicle can be broader earlier in their growth. Wiseman predicted that if this follicle-expansion idea was correct, then if he induced a juvenile to grow a second feather, the tip of the second feather would be more oblique than the tip of the first. Furthermore, this difference between original and induced rectrices should not exist in adults because they had already grown at least one expanded-follicle feather. As far as I know, neither Wiseman nor anyone else has actually done the analysis to test this prediction, but Wiseman's cardinal feathers did provide us with an early clue that induced feather growth could be influenced by environmental factors varying annually.

The central assumption of ptilochronology is that a reduction in growth bar width indicates a lowering of nutritional condition. Tom Waite and I (Grubb *et al.* 1991) examined Wiseman's feather collection with the idea of determining whether factors, such as day length or ambient temperature, not directly related to nutrition might also be related to feather growth rate. Wiseman had kept the feeders where he caught the birds supplied with food year-round, so we could assume that the birds had been adequately fed throughout the year.

We used for the analysis an original and induced tail feather from 186 cardinals. We inferred from our work with other songbird species that regeneration of the 10 growth bars from which we computed mean daily growth of an induced feather had occurred at about the middle of the 30-day period after the original feather had been pulled. Therefore, to assess the effects of day length and ambient temperature on feather growth, we first determined mean day length (minutes between sunrise and sunset) at 40° North Latitude and mean daily maximum, minimum, and average temperatures (°C) in the Cincinnati area for the 30-day period after each pull date. We used multivariate analyses to determine whether growth bar width, total feather length, or feather mass was related to age or sex of the birds, season of the year, photoperiod, or air temperature.

The most striking result was the relation of season with all three measures of feather growth (Figure 3.1). The cardinals tended to regenerate their tail

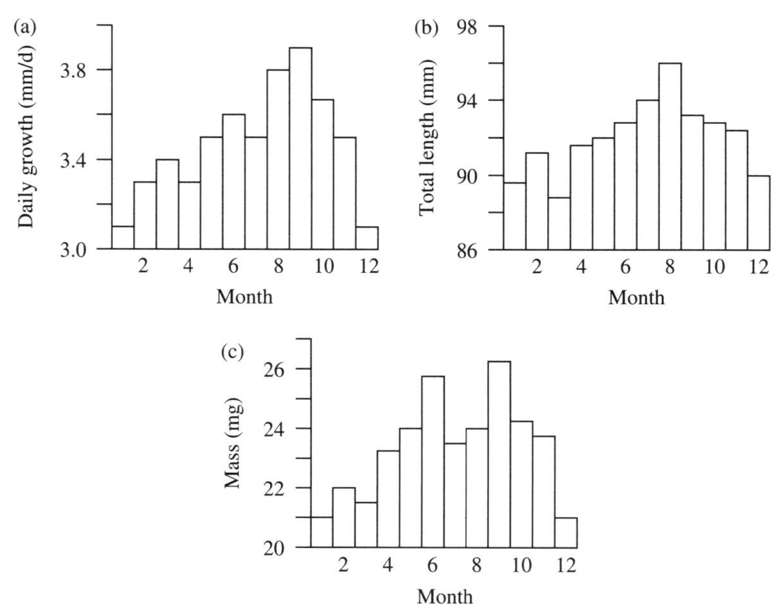

Figure 3.1 Mean regeneration of R6 rectrices in northern cardinals throughout the year (January to December). (a) width of daily growth bars on the induced rectrix, (b) total length of the induced feather, and (c) mass of the induced rectrix. From Grubb *et al.* (1991).

feather most slowly and to the smallest total length and mass during the winter months. Males tended to grow a longer and heavier induced rectrix than did females. The induced tail feathers grew at a faster daily rate and to a greater total length and mass at higher air temperatures.

Arthur Wiseman's feather collection enabled us to determine that several factors other than a bird's nutritional condition may influence the rate at which a feather is regenerated. Aside from differences that might have been expected due to age- and sex-specific variation in such factors as body size, social dominance (discussed later in detail), and foraging efficiency, there was a pronounced seasonal pattern in feather growth. All four age/sex categories of cardinals displayed an annual cycle with more rapid growth in summer than in winter. The results appear to limit the freedom with which ptilochronology can be used to study various aspects of avian behavior and ecology. The take-home message from the cardinal study is that variation in feather growth due to age, sex, or season of the year must be accounted for in the design of experiments.

In an experiment with American tree sparrows, Douglas White and Dale Kennedy pursued further the issue of non-nutritional control of feather-growth rates (White and Kennedy 1992). In earlier work (White *et al.* 1991), they had found that while female European starlings fed their nestlings, the birds grew an induced tail feather only about 75% as fast per day as they had grown the original feather during the previous molt. Two possible explanations came to mind to explain this discrepancy, either (1) hormonal factors were unfavorable for growing feathers during the breeding season, or (2) the great nutritional demands of parenthood during the breeding season prevented adequate resources from going to feather growth. They addressed these alternatives by exposing captive tree sparrows to either artificial short days (8 hours of light and 16 hours of darkness) that would prevent the onset of the reproductive cycle, or long days (20 hours of light and 4 hours of darkness) that are known to stimulate hormone production and consequent ovarian growth. Since all birds were provided with *ad libitum* food and water, only day length and the hormonal reaction it would provoke differed between the two groups of birds. White and Kennedy predicted that if reproductive hormones are sufficient to slow feather growth, birds on long days should produce induced feathers with narrower growth bars than birds exposed to short days. Conversely, if daily feather growth is not related to the production of reproductive hormones, then birds on long days should produce induced growth bars as long or longer (since they had more daylight hours to feed) than birds on short days.

Female sparrows were captured and first housed on short-day lengths. Then, in late February, 10 randomly selected birds were moved into long-day conditions, with another 10 remaining as controls under short days. Four days later, the original outermost right tail feather was plucked from all birds. After 42 days, the full-grown induced feather was pulled from each bird and at the end of the experiment, the ovaries of all birds were examined.

As expected, hormone production in birds exposed to long days greatly stimulated ovarian development. Thus, the researchers had a way of verifying that reproductive tendencies differed between the two groups.

Induced tail feathers of the long-day birds were significantly shorter and less massive than those of the short-day birds. However, growth bar width did not differ between the two groups. Thus, the long-day birds grew their feathers at the same daily rate as short-day birds, but grew them for fewer days. Based on these results, White and Kennedy suggested that the duration of rectrix growth, measured in days, is controlled hormonally while the rate of growth measured by growth bar width is controlled nutritionally. We need to remember that this conclusion applies only to the females they examined. In nature, of course, birds do not enjoy access to unlimited food as they did in this experiment, so that during the breeding season, we might expect to find narrower growth bars as a result of nutritional constraint. But this experiment clearly showed that some aspects of feather growth are functions of factors other than nutritional condition, in particular, seasonally varying hormone levels.

Elke De Ridder and her colleagues further supported the conclusion that hormone levels can influence feather growth (De Ridder *et al.* 2002). They had noticed large differences in the breeding-season behavior of male and female Belgian European starlings and wondered to what extent the variation was due to the higher testosterone levels of breeding males. They addressed their question by implanting packets of the male hormone, testosterone, into captive females of the species and implanting empty packets into other females as a control. The extra testosterone produced numerous behavioral and anatomical changes, but we will concentrate on its effect on the growth of induced feathers.

In Iate December, testosterone was implanted in eight females with seven females retained as controls. (We are not told whether the birds were assigned to the two groups at random, an important characteristic of any controlled experiment. Also, during the course of the study, fully seven of the 15 birds died, casting some doubt as to the validity of findings for the remaining animals.) At the time the birds were implanted, the two outermost tail feathers of each were plucked and stored. In late January, all the surviving birds were re-caught and the induced feathers plucked. For analysis, the measurements for right and left feathers were averaged.

While there was no difference in the total length of the original feathers of control and treatment birds, the total length of the induced feathers of testosterone-treated females was significantly shorter than that of the control females. Thus, we now have evidence that male hormones as well as female hormones (White and Kennedy 1992) can cause a shortening of induced feathers. Unfortunately, Ridder and colleagues did not present data on growth bar width of treatment and control birds. It would have been quite interesting to know if this study agreed with that of White and Kennedy in finding no hormonal effect on daily growth rate of feathers, only on number of days committed to growth. Incidentally, how the "decision" concerning how many days to grow a feather is made at the level of the follicle is totally unknown. These hormone studies reinforce the message of the cardinal study that for feather growth rates to be ascribed to nutritional condition, other potential causal factors should be controlled analytically or, better yet, methodologically with control and treatment groups.

3.2 *Non-nutritional response to cold temperature and wind*

Through the past several decades, my students and I have devoted much attention to the behavior and ecology of birds in winter. As detailed further along, we have used ptilochronology to investigate nutritional condition at that time of year with an eye to learning something about risk of mortality from starvation. Early on, we recognized we would have to deal with the possibility that simple chilling of a feather follicle in very cold weather could retard feather growth independent of nutrition. According to the Q_{10} Rule of physiology, the rate of metabolic processes slows down by about one half with each 10°C reduction in temperature. Since feather follicles are quite close to the outer surface of a bird, we reasoned that they might be physically chilled enough in the cold Ohio winter to cause retardation of feather growth. (Actually, I got this idea from realizing that my own whiskers grew less rapidly on winter days spent outdoors). Because such a chilling effect could potentially confound any interpretation of feather-growth rates of birds in winter, Gregg Zuberbier and I set out to test the possibility with white-breasted nuthatches (Zuberbier and Grubb 1992).

Gregg designed and built an electronics system that controlled temperature and fan-generated wind speed in nine chest-type freezers (Figure 3.2a). Each freezer unit contained a plastic-covered wind-tunnel cage with a propeller on one end (Figure 3.2b). Containers of food and water (or snow) were built flush with the bottom of the wind tunnel so they would not obstruct airflow. Gregg programmed the chambers to furnish a daily light-dark regime of 8 hours of light and 16 hours of darkness, about the situation in Ohio around the winter solstice. We randomly assigned three temperatures, +5, −5, and −15°C, and three wind speeds, 0.0, 0.5, and 1.0 m/sec, in such a fashion that each of the nine freezers had a unique combination of one temperature and one wind speed. Throughout the experiment, the temperature assigned to each freezer unit remained constant over the 24-hour period, but the fans were programmed to turn off during the dark period to simulate conditions inside the tree cavities where nuthatches roost at night.

After the nine male nuthatches were in place, one to a cage, we began an acclimation period. On the first day, all birds were exposed to their assigned wind velocities and to an air temperature of +5°C; on the second day, six of the nine were exposed to −5°C, and on the third day three of those six were dropped to −15°C. After the birds had spent one more day at the nine

(a)

(b)

Figure 3.2 (a) Gregg Zuberbier with freezer units, and (b) close-up of one freezer unit containing a wind-tunnel-cage used to determine induced feather growth in white-breasted nuthatches in response to variation in wind and temperature. Photographs by the author.

wind/temperature regimes, we plucked the outer right rectrix of all nine and returned them to their wind tunnels.

The prediction to be tested was that even if a nuthatch had abundant food, the combination of wind and cold temperature would be sufficient to retard tail feather growth. Therefore, throughout the feather regeneration period, we supplied the birds with *ad libitum* sunflower seeds and mealworms in the recessed containers. After eight weeks, we pulled the induced rectrices and released the birds at their capture sites.

We performed three trials of the experiment, designed to give us three replicates at each of the nine combinations of wind and temperature. However, the birds assigned to the −15°C, 0.5 m/sec treatment group in trials two and

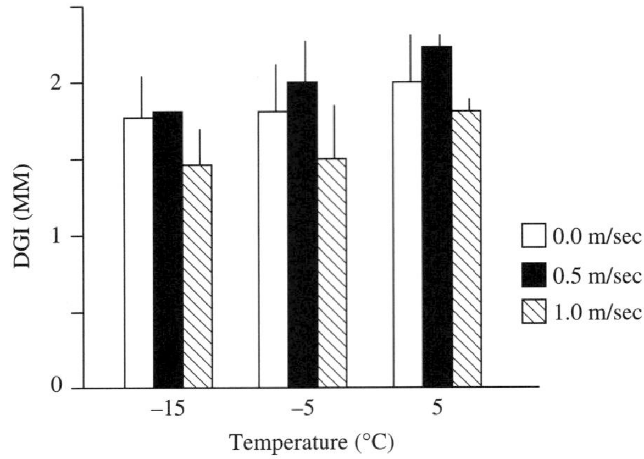

Figure 3.3 Mean width (± SD) of daily growth bars (DGI) on induced rectrices grown by white-breasted nuthatches under nine combinations of temperature and wind. N = 3 for all treatments except for −15°C, 0.5 m/sec, where N = 1. From Zuberbier and Grubb (1992).

three lost additional tail feathers during the experiment, so records for both were dropped from the analysis.

Results from a two-way analysis of variance on the total length of the original rectrix for all three trials indicated that the birds in all groups had come to the experiment equipped with about the same size original rectrices. Of central importance, there was no statistically significant effect of wind, temperature, or their statistical interaction on the total length, mass, or growth bar width of an induced feather (Figure 3.3).

These results had important implications for our winter fieldwork. They supported the view that cold temperature and wind do not affect the rate of feather growth independent of their possible effect on nutritional status. We then felt comfortable in attributing the causes of observed reductions in feather growth rate to factors other than direct effects of cold temperature and wind.

3.3 *Follicle history*

One possible variation of ptilochronology in monitoring nutritional condition is that sequentially induced feathers could be plucked from the same follicle. Such a procedure could be useful, say, in monitoring long-term change in habitat quality. Before we could proceed with such designs, however, we had to evaluate a serious potential confound. The scheme of using growth of sequential feathers taken from the same follicle as a long-term record of nutritional condition depends on the key assumption that the follicle does not change production rate of feathers simply as a function of producing several in sequence. That is, there is no effect of follicle history on feather growth independent of nutritional condition. To examine this assumption, Vladimir Pravosudov and I performed a controlled laboratory test of whether follicle history influenced the growth characteristics of a series of induced rectrices grown by house sparrows (Grubb and Pravosudov 1994a). We arranged for various randomly assigned sparrows to grow a first, second, or third induced rectrix from the outermost right follicle at the same time as they grew a first induced rectrix from the outermost left follicle. With time of year controlled by the design of the experiment, we reasoned that if there were no effect of follicle history, there should be no differences among the daily growth bar widths, total feather lengths, or total feather masses of first, second, or third rectrices induced at the same time on groups of randomly assigned birds.

We captured 12 male house sparrows in October, housed them individually in randomly assigned cages, and replicated the experiment twice, in 1991–92 and 1992–93. We randomly divided the birds into three groups of eight birds each (four birds each winter). For the eight birds in Group 1 we pulled the original R6 (outermost right) and L6 (outermost left) rectrices on November 1, and the first induced R6 and L6 rectrices on December 21 after the replacement feathers were fully grown. For the eight birds in Group 2, we pulled the original R6 on November 1, the first induced R6 and the original L6 on December 21, and second induced R6 and first induced L6 on February 13. Finally, for the eight birds in Group 3, we pulled the original R6 on November 1, the first induced R6 on December 21, the second induced R6 and the original L6 on February 13, and the third induced R6 and first induced L6 on April 9. The three successive intervals between pulling feathers were 51, 54, and 56 days in 1991–92, the last interval one day longer than in 1992–93 because of Leap Year. At each pull date, the birds' body masses were recorded to the nearest 0.1 gram. Figure 3.4 is a schematic of the plucking schedule, as a result of which we could determine whether the first, second, or third induced feather on one side of a sparrow's tail differed in growth characteristics from a simultaneously induced first induced feather from the other side of the same bird's tail.

Throughout their months in captivity, the sparrows were housed under the constant conditions of 22°C, 8 hours of light and 16 hours of darkness, and were maintained on *ad libitum* water and wild bird seed mix supplemented with fresh vegetables and hard-boiled eggs. At the end of each replicate, the birds were released. We discarded from the analysis two sparrows that lost additional rectrices during the course of the 1991–92 replicate, leaving us with eight, seven and seven birds, respectively, in groups 1–3.

Two salient results emerged from the experiment. Whether an induced R6 rectrix was the first, second, or third grown in succession from the same follicle, its daily growth, total length, and mass were all significantly positively related to the same values of the first induced L6 rectrix grown at the same time. Second, neither the first, second, nor third induced R6 rectrix grown from the same follicle differed from the mean value for the three induction dates in growth bar width, total length, or mass (Figure 3.5).

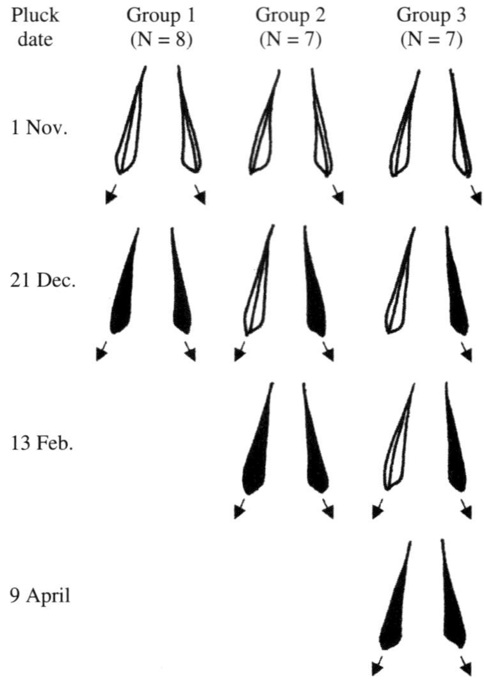

Figure 3.4 Plucking schedule in an experiment exploring effects of follicle history on feather growth (Grubb and Pravosudov 1994). The intent was to see if the first, second, and third induced rectrix from the same follicle differed in growth rate. For three randomly assigned groups of captive house sparrows, arrows indicate the dates over the course of the winter when original (white) and induced (black) R6 and L6 rectrices were pulled. Sample sizes are shown in parentheses.

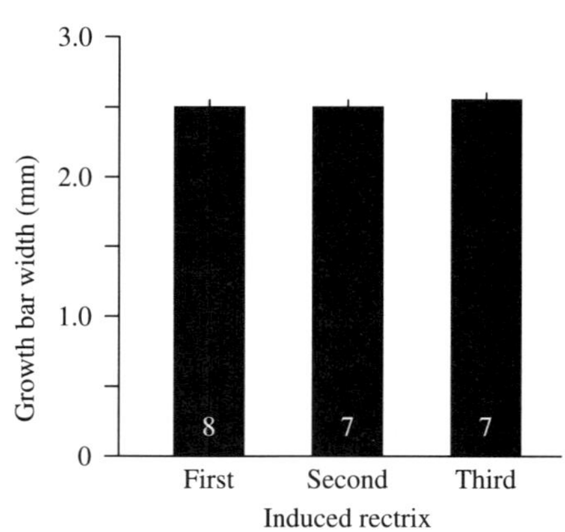

Figure 3.5 Mean (± SD) daily growth of male house sparrow first, second, or third R6 rectrices grown over a series of weeks from the same follicle. Sample sizes are shown in the bars. From Grubb and Pravosudov (1994).

A two-stage plucking schedule applied to free-ranging Spanish great tits produced similar results (Senar *et al.* 1998). First induced rectrices collected in January did not differ in growth bar width (2.35 ± 0.03 SE mm) from second induced rectrices collected from the same follicle in March (2.34 ± 0.03 SE mm).

Taken together, these results suggest that follicle history will not be an important confounding variable in any long-term study involving multiple feathers from the same follicle. However, we should still be cautious, especially until we have the results of studies using a longer series of feathers and/or shorter time intervals between inductions of successive feathers from the same follicle.

3.4 *Summary*

In Chapter 3, we have explored several variables that could confound interpretation of feather growth rates. The study of the annual cycle in northern cardinals and the two hormone-manipulation experiments all suggest that circannual rhythms and the hormonal cycles driving them must be accounted for. The best way of doing so is to have a firm grasp on when various original or induced feathers are grown and to control for time-of-year effects in manipulative work. There seem to be rich possibilities for exploring the effects of various hormones and hormone combinations on feather growth.

Work on peripheral wind-chill and follicle history in relation to feather growth has not demonstrated any effect. Nevertheless, it will be prudent to control for these variables methodologically in experimental work and analytically in descriptive studies.

To summarize Part I, ptilochronology has been directly related conceptually to the concept of biological fitness. Measures of induced feather growth have been put forward as fitness components of a subtle and early-warning variety. I have argued that induced feather growth is an unambiguous indicator of fitness and, as such, has advantages over more traditional measures such as body mass or percent body fat.

With the caveat that I am not a disinterested observer, I have responded to early criticisms of ptilochonology, particularly of some of its assumptions. Those early criticisms were heuristic in stimulating further work.

The latter sections of Part I probed for limitations to the conceptual scope of ptilochronology. We now know that proximate factors such as variation in day-length, with its influence on hormone levels, appears to have an effect on some components of feather growth independent of nutritional condition of the bird.

Finally, two examinations of additional potentially causal factors, cold temperature and wind, and follicle history, both failed to reveal any effects that might confound the main feather growth/nutritional condition nexus. However, it is important to remember that characteristics of these variables outside of those measured (i.e. colder, windier, feathers plucked closer together in time) could have an impact on feather growth rates.

Part II Applied Ptilochronology

Ptilochronology is now being employed in imaginative ways to explore the evolution, ecology, and conservation biology of birds. In the following chapters, I follow the avian annual cycle, more or less, in reviewing applications of ptilochronology to studies of habitat quality, food caching, social behavior, individual quality, reproductive effort, nestling condition, and prolonged brood-care.

4 *Habitat quality*

4.1 *Territory size*

The projects reviewed in this chapter take several approaches to the subject of habitat quality, but all focus on food intake rate as the measure of interest. Because of this focus, using feather growth rate was an attractive option. We begin with a project relating territory size to nutritional condition.

Loggerhead shrikes have become a species of concern in North America. Once relatively common, this bird has recently undergone a precipitous decline and is currently diminishing in numbers at a rate of about 5% per year. Shrikes are "perch and pounce" predators, requiring perches from which to hunt. With consolidation of farms and changing farm practices in recent decades, removal of fences, hedgerows, and "sentinel trees" in the middle of fields has vastly reduced potential hunting perches. Reuven Yosef and I studied the impact of reduced hunting sites on the nutritional condition of shrikes (Yosef and Grubb 1992).

Shrikes do not hunt on the wing. Therefore, we hypothesized that a territorial shrike requires some minimum amount of grassland or pasture within scanning distance from hunting perches, a portion of the territory we termed the utilizable area. Because the density of hunting perches is variable, we also hypothesized that in areas with few hunting perches, a shrike must defend a very large area overall in order to possess enough utilizable area, and that populations have been declining because continual removal of fence rows and other hunting perches has made utilizable areas too costly to defend due to the increased dead space. Although we tested four predictions from the hunting-perch hypothesis, I will focus on the one dealing with feather growth. Because of the increased amount of commuting among low-density hunting perches in large territories, the energy cost of flight should increase at the expense of energy available for feather growth. We tested this prediction with post-breeding permanent-resident shrikes at the MacArthur Agro-ecology Research Center, a 4,000-hectare cattle ranch that is a unit of the Archbold Biological Station, Lake Placid, Florida.

Territorial shrikes were captured in baited noose traps, color-banded, and released. We used the shrikes' aggressive responses to taxidermic mounts and playbacks of shrike vocalizations to map the borders of all territories on the ranch. During June, we pulled the R4 (right fourth) rectrix from territorial adult shrikes, recaptured the birds 5–6 weeks later, pulled out the fully grown induced rectrix, and calculated the feather's daily growth as described earlier.

We constructed a benefit-cost ratio of energy obtained to energy spent by dividing prey captures per hour of hunting from a perch by the percentage of time spent flying between perches. As capture rates per hour of hunting from

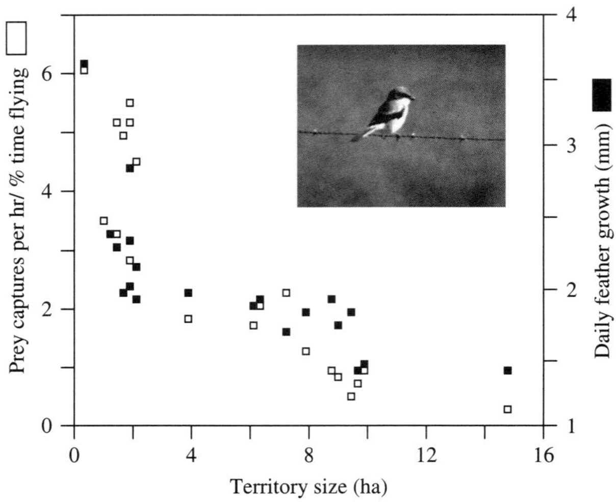

Figure 4.1 Relationships among a benefit-cost index of net energy intake (empty squares), daily feather growth as an index of nutritional condition (filled squares), and territory size in post-breeding loggerhead shrikes. From Yosef and Grubb (1992). Photograph by Reuven Yosef.

a perch were about the same for all territories, and as birds in larger territories had to fly more between perches, it was not surprising that the shrikes' benefit-cost ratio went down with increasing territory size. What was striking was how closely matched that ratio was by the decrease in daily feather growth (Figure 4.1). It appeared that the net energy intake indexed by our benefit-cost ratio explained very well the decrease in nutritional condition of birds on larger territories. In a subsequent study, Yosef and I showed that if we artificially increased the density of hunting perches on the MacArthur Ranch by "planting" fence posts across the landscape, average territory size went down and the density of birds per unit area increased (Yosef and Grubb 1995). Thus, we demonstrated a possible conservation measure for reversing the population decline of the species, at least in grassland or pasture habitats.

4.2 *Successional stage*

As a general rule, the vegetative composition of a site changes in a predictable fashion with time. For example, the first plants to colonize bare rock will be lichens and over a period of centuries or millennia, accumulation of organic matter may, in many parts of the world, lead to eventual progression toward forest. Such a phenomenon is called primary community succession because it is the first succession at the site, and the various communities of plants (and associated animals) through time are called successional stages or seral stages.

Secondary community succession occurs at sites where some agent has set back the community toward the beginning of succession, but not all the way.

For example, forest fires or logging set off a round of secondary succession. Management plans for many bird and other animal species sometimes consist of attempts to preserve certain successional stages of habitat. Ptilochronology has proven useful in determining the relative quality of successional stages for several bird species.

The pine woodlands of the American South have become intensively managed for paper pulp and lumber. Various management plans lead to different proportions of woodland in various age classes. Philip Stouffer and his students (Brown *et al.* 2002) investigated the quality of these various successional stages as habitat for hermit thrushes wintering in Louisiana woodlands. This thrush is one of a category of North American songbirds called short-distance migrants because they winter in the southern US rather than continuing farther into Central or South America.

The researchers divided their study woodlands into four categories, sapling-stage pine (5–9 years old; mean height 3 m), pole-stage pine (13–16 years old; mean height 5 m), mature pine (height range 15–20 m), and hardwood (oak, beech, and magnolia; height 15–25 m; Figure 4.2). They banded thrushes in the various community types and, among other variables, measured growth bars on the induced R6 rectrix of each bird using the standard technique detailed earlier.

Figure 4.2 The habitat categories, (a) sapling-stage pine, (b) pole-stage pine, (c) mature pine, and (d) hardwood, employed in a study of habitat quality for hermit thrushes wintering in Louisiana. Photographs by David R. Brown.

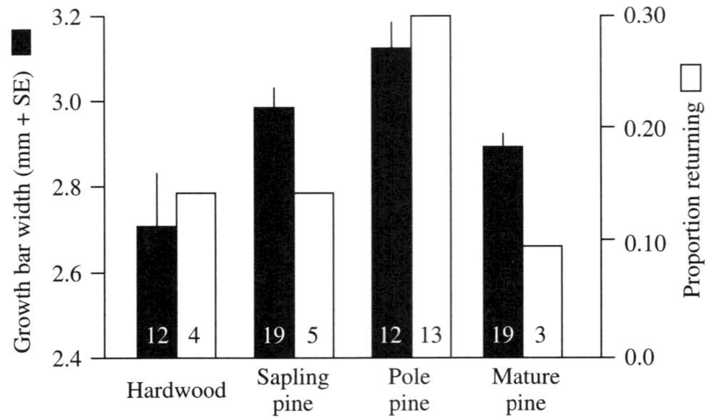

Figure 4.3 Daily induced feather growth in hermit thrushes wintering in four habitat types in Louisiana (open bars), and proportion of birds returning from one winter to the next in the same four habitat types (filled bars). Numbers of birds from which induced feathers were collected or which were recaptured the second winter are shown in the bars. From Brown *et al.* (2002).

Feathers in different habitats were re-grown at significantly different rates. Birds wintering in the pole-stage pine habitat grew out feathers significantly faster (3.107 ± 0.061 SD mm/day) than did those in hardwoods (2.821± 0.070 SD mm/day), with feathers of birds in sapling-stage and mature pine growing at intermediate rates (Figure 4.3). One could argue that feather growth was a sensitive indicator of nutritional condition, being lower in the hardwood stands than in any of the pine habitats, but subcutaneous fat levels were misleading as an index of nutritional condition as they did not differ among habitats. It is interesting that feather growth was highest in the pole-stage woodlands even though censusing and mist-netting showed that the thrushes were most dense there. Brown *et al.* ascribed the better nutritional condition of thrushes in pole-stage woodlands to the high density of fruits there, at least during the first part of the winter. The larger thrush territories and lower fruit densities in hardwood stands are reminiscent of the shrike territories in Florida, and the lower nutritional condition of the hardwood-dwelling thrushes may also be due to the higher energetic costs of commuting among sources of food.

This hermit thrush study provides us with the first chance to see whether nutritional condition indexed by feather-growth rate is really related to a major fitness component, namely survivorship. Because Brown *et al.* mist-netted birds at both the beginning and end of the winter, they could calculate within-winter survivorship in the four habitat types. Across all four habitats, such survivorship was about the same; between 60.7% and 63.6% of birds captured at the start of the winter were recaptured at the end of the winter. Thus, it appears that the increased nutritional condition in pole-stage stands indicated by increased rate of feather growth did not correspond to increased within-winter survivorship there. However, feather-growth rates were positively related to return rates from one winter to the next. Pole-stage pine, the habitat with the highest rate of feather growth, had more than twice the total number and proportion of between-year returns as any other habitat (Figure 4.3). It seems possible that even though birds in other habitats with lower feather growth rates did not die during the winter, they left for spring migration in poor enough nutritional condition to cause mortality sometime between leaving and returning to the wintering habitat the next autumn.

This result points out the inherent difficulty of interpreting annual mortality records for migrant species. In the case of these thrushes, we cannot know what happened to them during the time they were away from the wintering ground. A more conclusive examination of the relationship between feather growth rates and mortality will come shortly when we examine these two factors in permanent-resident birds monitored in one place year-round.

It is tempting to conclude as a management policy for wintering hermit thrushes that pole-stage habitat should be widely distributed over a landscape. This is probably a good idea in any case, but a possible confounding factor needs to be considered before the correlation between pole-stage pine and enhanced nutritional condition may be considered cause and effect. In their periodic censusing, Brown *et al.* found that the earliest arriving birds in the fall settled preferentially in the pole-stage pine habitat and that later-arriving birds scattered out among the other three habitat types. Thus, we cannot rule out the possibility that birds in the pole-stage sites grew tail feathers the fastest because they were of the highest-quality birds to begin with. Their higher quality had made it possible for them to migrate the fastest and arrive first. A critical test would be to remove early-arriving birds from pole-stage areas so that later-arriving birds could find room there. If these later-arriving birds also grew rectrices at the fastest daily rate in pole-stage plots, we could have more confidence that the effect was due to the quality of the successional stage and not the quality of the bird.

As long as we are discussing experimental design problems, there is another that illustrates the difficulty of doing controlled experiments in nature. In this hermit thrush project, it is possible (though unlikely) that feather growth rate was actually a function of something about location other than forest stage. For example, suppose fruit density was actually a function of soil type and it just so happened that a disproportionate number of pole-stage plots were on the favorable soil. If this experiment were being done with bean sprouts in petri dishes, we would randomly assign sprout height to individual dishes, but to follow such a design in nature would take at least two decades of forest growth before the experiment could begin. Some things are just not possible in the real world.

At the Archbold Biological Station in central Florida, Glen Woolfenden and his colleagues continue to build an impressive long-term demographic study of Florida scrub-jays (e.g. Woolfenden and Fitzpatrick 1990, 1996). An important food and cover type for the species is oak scrub, a combination of several species of stunted oak trees. The oak scrub community is maintained by periodic fire in a kind of permanent successional stage sometimes called a fire disclimax. Not all of the scrub-jay territories are completely covered by oak scrub and Breininger *et al.* (1995) found that production of young was positively correlated with the amount of oak scrub in a territory. Woolfenden, John Fitzpatrick, and I collaborated on a ptilochronology study aimed at assessing the impact of oak scrub on the nutritional condition of fledgling scrub-jays (Grubb *et al.* 1998).

Woolfenden and Fitzpatrick (1990) and Mumme (1992) had concluded that predation was the dominant cause of mortality in young scrub-jays throughout the period between the time they hatched and the time after fledging that they became able to find food on their own. If fledgling mortality is caused by predation and is relatively low in open oak scrub, then two complementary mechanisms might be responsible. First, independent of the nutritional condition of the fledglings, oak scrub may reduce predation either by making the young birds harder for predators to capture or, since the ground cover is sparse under oaks, by making predators easier for the jays to detect. Second, oak scrub may contain abundant food, permitting high feeding rates of nestlings and fledglings and therefore reducing the loud begging vocalizations of young birds thought to attract predators (Mumme 1992). If mortality is lower in open oak-dominated territories only because of lower predation, then nutritional condition should be independent of the amount of oak scrub in the territory. If mortality is lower in open oak-dominated territories because either such territories contain more food or the better-fed fledglings there are less conspicuous, or both, then nutritional condition should be positively correlated with extent of oak cover. In order to test these two predictions, Woolfenden, Fitzpatrick, and I employed ptilochronology to index nutritional condition of fledglings in territories differing in extent of open oak scrub.

We addressed three questions: (1) is growth bar width a valid measure of nutritional condition in fledgling scrub-jays, (2) do factors other than territory quality cause variation in nutritional condition among fledglings in different territories, and (3) is nutritional condition of fledglings positively related either to amount of oak scrub or to its proportional representation in a territory.

As he does every year, Woolfenden banded nestling scrub-jays and then recaptured and color-banded surviving fledglings about 80 days after fledging had occurred (Figure 4.4). Young leave the nest when their rectrices are only 1–2 centimeter long, but by about 80 days later, when Woolfenden plucked the outermost right rectrix, it was fully grown. We used feather growth during the period between the fledging date and 80 days thereafter as a measure of nutritional condition of the newly fledged jays. Woolfenden shipped the feathers to me identified only by code and I measured them "blind" in the usual fashion.

Brood sizes at the time of plucking ranged from one to four. For all analyses, we averaged the feather measurements for all the young jays surviving in each territory's brood each year.

Figure 4.4 Fledgling Florida scrub-jay. Growth bars are quite evident on this bird's tail feathers. Photograph by the author.

It is known that in scrub-jays a positive relationship exists between nestling size and fledgling survivorship (McGowan 1987; Fitzpatrick *et al.* 1988), which leads to the conclusion that nestling body size is positively related to nutritional condition and fitness. Growth bar width and feather mass were positively and significantly related to nestling day-11 body mass. A similar trend held for total feather length, but fell short of significance. From these results, we concluded that growth bar width, feather mass, and probably feather length were valid indices of nutritional condition in the young newly fledged jays.

Using analysis of covariance (ANCOVA), we checked to see if any other factor Woolfenden measures every year was related to feather-growth indices. For each territory, this analysis accounted for (1) day of year when a clutch was first incubated, (2) number of fledglings, (3) number of non-breeding adult jay helpers, (4) average body size (tarsus length) of the parents, (5) year of the project (1990–94), (6) whether the pair-bond of the breeding pair was new or continuing from previous years, and (7) the identity of the individual territory, which remained constant over the five years of the project. Of the variables tested, only territory was significantly related to growth bar width. Year, pair-bond, and territory were all significantly related to total length and mass of the rectrix. Rectrices of fledglings were longest in 1992 and shortest in 1991. Rectrices of fledglings in territories held by new pairs were shorter and lighter in mass than those grown in territories held by pairs with a pair-bond continuing from previous years.

We next inquired whether amount of open oak scrub could be a measure of habitat quality responsible for the significant variation among territories in nutritional condition of fledglings. We looked for relationships between feather characteristics and both the total area of oak in a territory and the proportion of a territory in oak cover. Because we had found relationships between indices of feather growth and both year of the study and length of the parents' pair-bond, we included those two variables in this second analysis.

No measure of feather growth was significantly related to absolute area of open oak scrub in a territory. However, the relationships of all three feather measures with proportion of oak cover were positive when the absolute area of oak had been controlled statistically. While the relationship between proportion of open oak scrub and total length of feather was significant, those of feather growth bar width and feather mass trended the same way but not quite statistically significantly. These results lend some support to the idea that even within the relatively uniform Florida oak scrub, habitat composition of territories affects nutritional condition of fledgling jays. Moreover, body growth rate as a young nestling is correlated with subsequent survival (McGowan and Woolfenden 1990), and our study showed that feather growth rate after fledging is correlated with nestling mass. Together, these results imply that nutritional condition of both nestlings and fledglings affects first-year survival. In any case, our findings show that subtle differences in habitat quality—specifically the proportional representation of oak scrub within a territory—may help determine overall reproductive success. We concluded that territories containing a high proportion of oak cover permit more efficient foraging and reduce the energetic cost of territorial patrolling compared to those with lower proportions of oak.

4.3 *Moisture*

Two studies have used ptilochronology to demonstrate variation in nutritional condition of birds occupying habitats along a moisture gradient. Roberto Carbonell and J. L. Telleria introduced their study of a small European songbird, the blackcap, by noting that many Old World passerines are decreasing in abundance in the Mediterranean region, apparently due to drought (Carbonell and Telleria 1999). Noting that blackcaps molt their feathers in summer under very dry conditions, they reasoned that if nutritional condition of the birds varied with environmental moisture, across three locations along a north-south moisture gradient in Spain, birds in the most moist north should grow feathers fastest, those in central Spain should be intermediate, and those in the south should show the slowest daily growth. Rainfall from April to August in the three locations, respectively, was 350, 270, and 130 mm. The assumption behind this idea was that moisture should control rate of growth of plants that, in turn, should control the rate of population growth of the birds' insect food supply.

Birds were mist-netted between mid-May and the end of July, the fifth left and right rectrices were plucked, and the birds were released. Importantly, the birds were not caught again, so the paper's analysis is based entirely on original feathers, a matter I will return to.

Results were equivocal. There was no difference among the three populations in either width of growth bars or total feather length, indicating that birds in all three areas grew their feathers at about the same rate per day and for the same number of days. However, there was a difference in the mass of feathers, those in the south being lightest. Juveniles did have narrower growth bars than adults in the two southern populations, but not in the northern one.

Carbonell and Telleria interpret their results as indicating blackcaps in the south were more nutritionally stressed, presumably due to the overall drier climate. They point out that juveniles began growing their tail feathers in the nest and that some of those nests in the south would have been in quite dry habitat. By contrast, they say, following the breeding season, adults molt their rectrices only after having moved to moister areas.

Because the study was based on original feathers, some ambiguity appears to have entered the analysis. According to the authors, some second-year blackcaps, which were counted as adults, were still carrying juvenile rectrices. If the proportions of such birds differed among the three areas, interpretation of results would have been confounded. Nevertheless, this study did present some evidence supporting the notion that environmental moisture (and temperature) can play a role in controlling the nutritional condition of birds.

Over the past decade or so, the continuing decrease in numbers of Neotropical migrant birds breeding in North America has called forth a variety of studies searching for causes of the declines. Part of the recent effort has focused on the southern wintering grounds, where many migratory bird species actually spend more of the year than they do in the breeding range. One such project, by Allan Strong and Thomas Sherry, involved intensive ecological study of ovenbirds on the island of Jamaica (Strong and Sherry 2000).

Just as they do in the breeding range, wintering ovenbirds feed on the ground and, as it turns out, largely on ants. Birds and their ant prey were monitored in three distinct habitat types, shade-tree coffee plantations, second-growth scrub, and undisturbed dry limestone woodland, all sites located close together. Ovenbirds were caught and a rectrix plucked early in the winter, then the birds were recaptured and the induced feather retrieved in February and March, shortly before the birds departed northward (Figure 4.5).

Growth rates of induced rectrices varied across the three habitats and were significantly slower in the driest location, the dry limestone woodland, than in the other two environments. On an individual-bird basis, there was a good correlation between the total biomass of ants in a bird's territory at the start of the winter and the daily growth of that bird's induced tail feather (Figure 4.6). Thus, the study was able to tie a known reduction in the food supply (ants) to a known reduction in the feather growth rate index of nutritional condition. Records were not sufficient to explore the relationship between nutritional condition and between-winter recapture rates, so any relation between feather growth rate and annual survivorship remains unknown.

Figure 4.5 Jamaican (a) second-growth scrub, and (b) shade-tree coffee plantation in which the nutritional condition of (c) ovenbirds was studied. In (d), growth bars are particularly noticeable on the underside of the outermost left (L6) rectrix of this ovenbird. Photographs by David R. Brown.

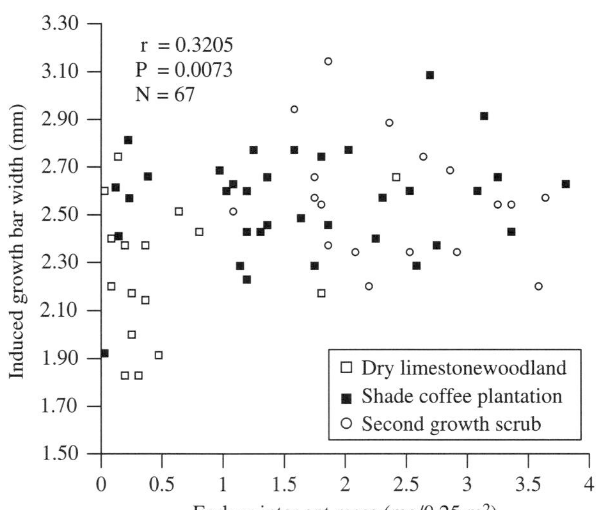

Figure 4.6 Relationship between early-winter biomass of ants in the home-ranges of ovenbirds wintering in Jamaica and growth rate of the birds' induced R3 rectrix. From Strong and Sherry (2000).

In contrast to feather growth rate, fat supply appeared to be inversely related to nutritional condition. Birds in dry limestone woodland had the lowest food supply, but the highest fat scores, the latter a measure of the extent of their fat supply. Such a result is reminiscent of other projects finding that birds with a less predictable food supply carried with them a higher fat "insurance policy" (Rogers 1987). This project clearly indicates that we should look further to the wintering grounds of Neotropical migrants for clues to causes of recent population declines on their breeding grounds.

4.4 *Anthropogenic modifications*

We complete this section on habitat quality with six studies, all focused on the impact of humankind's modification of habitat. Clearly, such information will be useful in constructing avian management plans. The first example stems from another portion of Reuven Yosef's and my project on Florida loggerhead shrikes. In addition to holding territories at the Archbold Biological Station, these shrikes occupy a variety of artificial and more natural habitats in central Florida. We monitored feather growth with an eye to assessing the impact of various habitat types on shrike nutrition (Grubb and Yosef 1994).

During March and April, we live-trapped shrikes along 524 km of roadway in five central Florida counties. The habitat in the vicinity of each capture site was recorded as (1) built-up urban, (2) palmetto scrub, (3) citrus, or (4) fenced pasture. At every capture site, the habitat was the same on both sides of the road. Before releasing each bird at the capture site, we determined age and sex, recorded tarsus length (more formally, the length of the tarsometatarsus bone of the leg), and pulled and stored the right fourth rectrix. Growth bars are most easily discerned on feathers that are not pied or marked by pigment bars. While many other projects have used the outermost (sixth) rectrix, in loggerhead shrikes the fourth is the outermost rectrix that is predominantly monochromatic. Feathers were coded by Yosef so that I could measure them in a "blind" fashion without knowing from which of the four habitat types they had been taken.

We restricted the analysis to feathers from birds known to be adults at the time of capture. Shrikes in southern Florida appear to spend their entire adult lives on the same territory (Yosef 1992). Therefore, we assumed that the (original) feathers we collected from adults had been grown at the same site during the molting period the previous autumn. As juvenile birds born the previous summer could have grown their rectrix in a different habitat type before dispersing to the site where we caught them, we omitted their rectrices from the analysis.

A multivariate analysis revealed that the total length of the R4 rectrix was not related statistically to tarsus length, year of capture, sex of the bird, or habitat type. By contrast, growth bar width and total mass of the feather were both significantly related to habitat type, but not to tarsus length or year. Daily growth, but not feather mass, was related significantly to sex, with

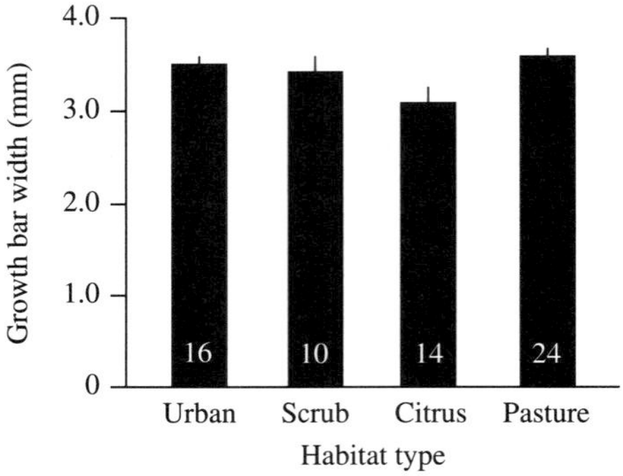

Figure 4.7 Mean ± SD daily growth of the original R4 rectrix of territorial adult loggerhead shrikes in four habitat types in south-central Florida. The values for citrus and pasture were significantly less and greater, respectively, than the mean value for the four habitat types. Sample sizes are shown in the bars. From Grubb and Yosef (1994).

males growing feathers faster. The rectrices of birds in citrus groves grew significantly less each day and were less massive when fully grown than the average feather in all four habitat types. By contrast, rectrices from shrikes in pastures were longer, heavier, and had wider growth bars than the average for the four habitat types. In all three feather characteristics, feathers from urban and scrub habitats were intermediate between those from citrus and pasture (Figure 4.7).

It seemed that something about citrus and pasture, respectively, had been bad and good for the nutrition of shrikes during the fall molting period. However, again we must realize that these are only correlations. As we have seen in other studies, it remains possible that the highest and lowest quality birds, respectively, were found in pasture and citrus in the first place, so the differences in feather growth could have been a function of bird quality, not habitat quality. Assuming, though, that the rate of feather growth in pasture and citrus was affected by habitat type rather than bird quality, its causation could have been direct, indirect, or both. In central Florida, shrike populations are densest in pastures, which suggests that short grass and abundant fence-post hunting perches together provide superior foraging habitat and consequent excellent nutrition for the species (Yosef 1992).

Florida citrus is routinely sprayed during the autumn with mitacide/insecticide compounds known to be toxic to non-target organisms (Hayes and Laws 1991). Such chemicals could have had a chronic sublethal effect on shrikes sufficient to retard feather growth. Alternatively, application of miticides/insecticides in groves could have reduced the shrikes' food supply, thereby reducing the birds' nutritional condition sufficiently to be detectable using ptilochronology. Also, it seems possible that nutritional condition could have been lowered by both direct and indirect effects of citrus-grove management practices. In any case, it seems clear that feeding rates (Yosef and Grubb

1992), induced feather growth, survivorship, and fecundity of shrikes living in citrus groves should receive additional attention.

A short time after our Florida study, Yosef used essentially the same methods to assess quality of two highly endangered habitats in Eilat, Israel (Yosef 1997). Eilat, at the northern end of the Red Sea, has been an important migration stop for birds migrating between Africa and Eurasia. Yosef wanted to demonstrate to local authorities that the remaining habitat patches of salt marsh and canal-side reed beds provide stop-over sites for migrants. To do this, he chose to examine the nutritional condition of a permanent-resident species, the graceful warbler, found in both the salt marsh and the reeds. As before, the analysis was based on growth characteristics of original feathers.

Again, only adult birds were used, the assumption being that they had grown the plucked original feather in the same habitat where they were captured. Although it is known that salt marsh is the native habitat of this warbler at Eilat, birds living in the reeds along a canal had original feathers with significantly wider growth bars (2.33 ± 0.19 SD mm versus $2.25 + 1.57$ SD mm) even though birds at the two sites were the same structural size. Yosef explained the rather unanticipated result, not as the salt marsh being deleterious to the warblers' nutrition, but as the reed-bed habitat being extra-rich in resources. Apparently, organic waste dumped into the canal from several sources had created a rich environment for algae-eating flies that, in turn, were an important food item for the warbler. Yosef went on to argue that because of the scarcity of remaining migration stops in the region, both the marsh and the canal habitat should be retained as undeveloped sites.

Two studies have used ptilochronology in an effort to understand keys to quality woodpecker habitat. Allan Carlson investigated the case of the endangered white-backed woodpecker in Sweden (Carlson 1998). During the second half of the twentieth century, this species underwent a sharp reduction in numbers across Scandinavia and Finland. The species feeds primarily on wood-boring and bark-inhabiting insects of dead trees and dead parts of live deciduous trees. Carlson reported that dead trunks formed 20% of stems in an average forest during the nineteenth century, but that by the late twentieth century, their prevalence had shrunk to about 1%. Similarly, at the end of the nineteenth century about 30% of trees in Swedish forests were deciduous, the rest being conifers of various species. By contrast, in the highly managed plantation-forests of today, only about 2–4% of trees are deciduous. Thus, there were good grounds for thinking that white-backed woodpeckers were approaching extinction as their food base in dead stems and deciduous trees became very scarce. Carlson reasoned that if this was the cause of the decline, in the few woodpeckers remaining today, he should see width of feather growth bars directly related to amount of dead wood, either as dead trees or as deciduous trees that he knew often carried dead branches. Furthermore, he predicted that growth bars on the feathers of museum specimens collected well before the character of northern forests began to change should be wider than those of birds sampled recently.

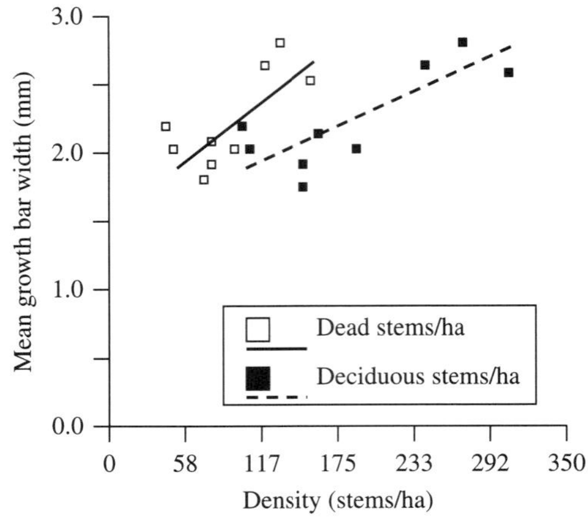

Figure 4.8 Relationships between two measures of territory quality, numbers of dead stems and deciduous stems per ha, and width of growth bars on the R4 rectrix of white-backed woodpeckers, a species in decline in Sweden. From Carlson (1998).

During the spring, Carlson captured breeding adult woodpeckers, pulled and stored the right fourth rectrix, and released the birds. He used only original feathers for his analysis and assumed that because adults of the species show a high degree of site faithfulness, the feathers sampled had been grown at the breeding site during the previous autumn's molt. He measured growth bars in the usual fashion and also did so on eight museum specimens collected between 1832 and 1942. In the territory of each living bird sampled, he tabulated the density (stems per hectare) of both dead trees and deciduous trees.

Carlson's results showed significant positive regressions between the width of feather growth bars and the densities of both dead stems and deciduous trees (Figure 4.8). Furthermore the growth bars on the R4 rectrices of the old museum skins (3.71 ± 0.41 SD mm) were much wider than on the living birds (2.20 ± 0.36 SD mm). Thus, Swedish ptilochronology supported the suspected nutritional basis for the species' decline in numbers throughout its range.

Carlson's project also highlighted the potential use of museum specimens for comparing environmental conditions through time. Unlike any other measure of nutritional condition, the growth characteristics of feathers as indexed by growth bar width are on display permanently and without variation. Furthermore, these records can be taken without damaging specimens. The potential for documenting changing environmental conditions using this comparative method is very great considering the hundreds of thousands of prepared bird "skins" from all over the world now lying in museum trays.

Carlson reminds us that once again we are dealing with a correlation. It is possible that birds with more deciduous growth in their territories were of higher quality to begin with. He did feel this circumstance was unlikely as the museum specimens with longer wings, indicating larger body size, did not have a rectrix with wider growth bars. Thus, the idea that larger, more socially dominant birds could command higher quality habitat was not supported.

Carlson concludes his account by noting that as a result of Northern European industrialized forestry, today, old-growth deciduous woodland is extremely scarce. He goes on to state that his feather growth results support the view that the observed decline of white-backed woodpeckers in Sweden is caused by deterioration of the forest landscape.

Official designation of the red-cockaded woodpecker as a federally endangered species in the US fostered a large number of projects throughout the Southeast focused on understanding the species' biology. As part of their long-term study in east Texas, Richard Conner and his colleagues examined growth rates of feathers in two types of pine habitat (Schaefer *et al.* 2005). In the eastern Texas part of the species' range, family groups occur in groves of longleaf pine and also in groves composed of a mixture of loblolly and shortleaf pine. These pinewoods are intensively managed for timber and paper pulp. Any difference in nutritional condition of the birds between the two woodland types could well influence management for woodland type.

Schaefer *et al.* knew red-cockadeds tended to avoid pine groves with an admixture of deciduous shrubs and small trees, so much so, that periodic fires, which favor pine trees over deciduous vegetation, are now a recommended component of management for the woodpecker. In their area, longleaf pine woodlands had been subject to more frequent burns and, thus, had less deciduous growth than the loblolly/shortleaf groves.

The logic for the project with red-cockadeds stemmed from work that David Peters and I had published earlier on the foraging behavior of downy woodpeckers, a close relative of red-cockadeds (Peters and Grubb 1983). We knew that in downy woodpeckers, there is a division of foraging locations between the sexes; males are usually found in higher parts of trees and on thinner branches than females. (I am fond of the "magic trick" of identifying to students the sex of downy woodpeckers with the naked eye at very great distance based solely on how high in a tree they are foraging.) Peters and I were interested in the cause of such sex-specific feeding. One possibility was that males and females were genetically programmed to forage at different heights. However, we knew that males were socially dominant to females, so a second possibility was that males were using their social dominance to exclude females from the upper regions of woodlands, perhaps because the food supply was richer there, the risk of predation less, or both.

Peters and I performed a removal experiment to test these possibilities. If males were excluding females from the preferred upper parts of woodlands, we reasoned that removing males would result in females foraging higher in the trees. However, removing females should not result in a so-called niche shift by males to lower parts of the forest as they could have used their social dominance to forage there in the presence of females had they wanted to. By contrast, if sex-specific foraging sites were genetically fixed, removing birds of one sex should have no effect on the foraging sites of the other.

The results of our experiment were clear-cut. When we removed males, females moved up in the woodland, but when we removed females, the males stayed where they had been. We concluded that in downy woodpeckers, at least, the sex difference in foraging is a consequence of the social dominance of males.

The Texans knew about our downy woodpecker experiment. They also knew that male red-cockadeds foraged higher in the forest than females. In fact, males often could forage above the level of the deciduous growth in loblolly/shortleaf pine groves far more than could the females. They assumed that females were forced by male dominance to forage within the mid-story, less-preferred deciduous growth or, if they ventured upward, they would come into greater competition with males. Thus, they hypothesized that between the low-deciduous longleaf pine and the high-deciduous loblolly/shortleaf pine, growth bars on the feathers of females should differ more than growth bars on males.

The researchers captured woodpeckers in their roosting cavities, weighed them, and pulled the second outermost right rectrix (R5) grown during the previous autumn molt. Thus, there was time variation in indices of nutritional condition between information on feathers grown during the previous molt and records of body mass applying to the date the feather was pulled, weeks or months later. The average width of six growth bars was calculated and the total length of the feather was also recorded. Only adults were used in the analysis and the sexes were treated separately.

Males did not differ between the two forest types in either growth bar width or feather length, but in the loblolly/shortleaf habitat were significantly heavier (Figure 4.9). Females from the loblolly/shortleaf pine forests had wider growth bars, longer feathers, and were heavier than their counterparts from the longleaf woodlands (Figure 4.9). It is worth noting in connection

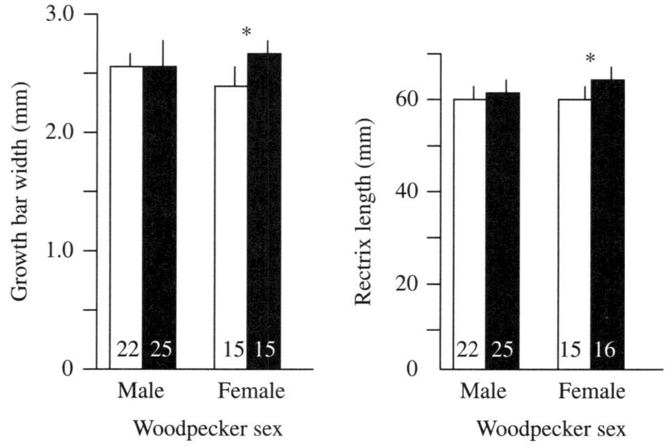

Figure 4.9 Mean ± SD growth bar width and total length of the R5 rectrix of red-cockaded woodpeckers in longleaf (open bars) and loblolly-shortleaf (filled bars) pine woodlands. Asterisks indicate significant differences, and sample sizes are shown in the bars. From Schaefer *et al.* (2005).

with the ambiguity of body mass records noted in the first chapter, that here the feather growth data suggest that the birds in loblolly/shortleaf were heavier because they were in better nutritional condition, not worse.

The feather growth data turned the working hypothesis on its head; instead of females in loblolly/shortleaf being in worse nutritional condition, as hypothesized, they were in better. Never mind; a good scientist can always come up with a new hypothesis to explain an unanticipated result. The difference in the adverse effects of foraging in longleaf for males and females is congruent with the idea that males selected the better location to forage in both habitat types. But why might females do better in the habitat with more deciduous growth? First, it turns out that the canopy in the loblolly/shortleaf was considerably taller than in the longleaf (Rudolph *et al.* 2002), so females in the former may have been more able to forage above the deciduous layer without impinging on the males' uppermost foraging zone. Second, there is evidence that the food supply might have been greater in the loblolly/shortleaf. The densities of dead and dying trees and of trees showing activity of the southern pine beetle (*Dentroctonus frontalis*) were significantly higher in the loblolly/shortleaf than in the longleaf. Red-cockadeds often concentrate their foraging activity on dying trees with high populations of various arthropods (Schaefer 1996). The authors argue that the increased prey base in loblolly/shortleaf may have been more than enough to reverse any negative effects of the deciduous mid-story, resulting in better nutritional condition for the woodpeckers in that forest type.

This study reinforces the importance of biological research to conservation plans. Without the work of Schaefer and his colleagues, forest managers might have assumed they should always favor longleaf pine plantations to aid recovery of this endangered woodpecker. We have seen evidence that that procedure may be precisely the wrong thing to do.

Finally, here we have a case where a rather important management decision depends at this point on only a correlation, not a known cause-and-effect relationship furnished by a manipulative experiment. Could it be, for example, that the reason red-cockadeds are heavier and females have longer rectrices with wider growth bars in loblolly/shortleaf is because the birds there are larger or of higher quality to begin with? This study used only original feathers and no correction was made for structural size. Of course, a paradox about endangered species is that in just those cases where the findings of manipulative experiments could have the most practical importance, such manipulations are prohibited by the very law applying endangered status.

Next, we turn to a ptilochronology study of human-generated fragmentation of the Amazonian rainforest. Brazilian rainforest continues to be rapidly turned into pastureland and soybean fields. A well known and very ambitious study has documented the reduction in bird species in forest patches of 1, 10, and 100 hectares left during clear-cutting operations (Bierregaard and Lovejoy 1989). Several rainforest bird species persisted even in the 1-hectare patches, an area less than 25% of the territory size of the more common Amazonian

Figure 4.10 (a) Corner of an Amazonian forest fragment, (b) wedge-billed woodcreeper, and (c, d) white-crowned manakin from a study of avian nutritional responses to fragmentation of tropical rainforest. Photographs by Philip Stouffer.

forest birds (Stouffer and Bierregaard 1995). Jeffrey Stratford and Philip Stouffer used feather growth rates of feathers collected in 1991 and 1992 to determine if birds in small patches were adversely affected nutritionally (Stratford and Stouffer 2001). They focused on two common species, the wedge-billed woodcreeper and the white-crowned manakin (Figure 4.10). The woodcreeper is very common in wet Neotropical forests, moves over tree bark looking for insect prey, and often forages in pairs within mixed-species flocks. The manakin forages for fruit and insects in the forest understory.

Birds of the two species were mist-netted in woodland fragments of the three size classes and in a large continuous forest nearby termed the reference area, were banded, had a feather pulled, and were then released. Only some of the birds were re-caught to provide an induced feather so the analysis combined, in an unclear fashion, measurements on original and induced feathers.

Stratford and Stouffer were able to show statistically that the feathers from the birds captured in fragments had narrower growth bars than those from birds in continuous forest (Figure 4.11). In the woodcreeper, there even seemed to be a cross-fragment trend toward slowest feather growth in the smallest-sized fragments.

How might fragmentation have affected feather growth and therefore presumably the nutritional condition of these permanent-resident birds? The

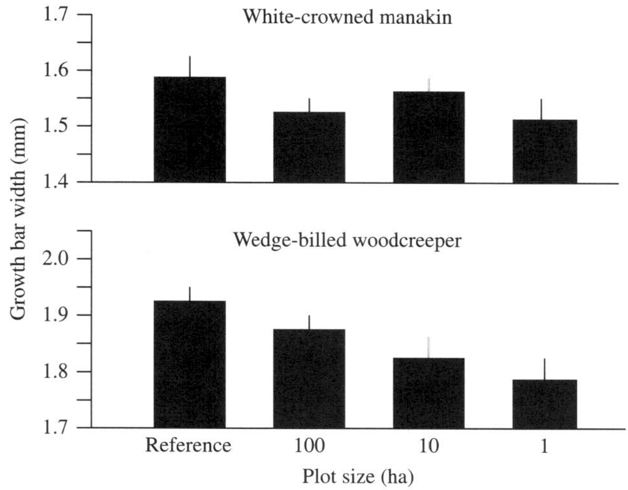

Figure 4.11 Mean (± SE) daily feather growth rates of two Amazon rainforest bird species in habitat patches of different size. The "reference" plot is continuous rainforest. From Stratford and Stouffer (2001).

researchers offer several possibilities. One is that through some unknown mechanism, the higher temperatures and lower humidity in the fragments directly retard feather growth. They suggest that for the woodcreeper, the collapse of mixed-species foraging flocks in fragments may have disrupted normal feeding efficiency, leading to reduced net energy income and, therefore, poorer nutritional condition. They note that insect abundance appears to decrease in small fragments, but that the fruit-eating manakin could actually benefit from the increase in the fruit supply along forest edges of even small areas.

Before leaving this study, we should point out some difficulties. Stratford and Stouffer suggest that some birds could have actually grown feathers in continuous forest and then dispersed to fragments (and presumably vice versa). Perhaps, a more severe problem stems from lumping measurements of some original feathers grown during the previous molt with those of other feathers induced in the fragment where the original was plucked, and in proportions that varied among the fragments and continuous forest. Original feathers were included to make up for an insufficient number of induced feathers from recaptured birds. We know nothing about the annual cycle of feather growth in tropical birds, but the results from the study of Cincinnati cardinals detailed earlier in this account urge caution in interpreting the present results.

The authors conclude by stating that their feather growth results suggest that fragments were inferior habitat for these two tropical forest songbirds. They point out, as has been done before in this account, that the next step is to see whether such reductions in nutritional condition relate to reductions in lifetime reproductive success. As a final note, we must congratulate the formulators of this fragmentation project for thinking big. Knowing that the rain forest was going to be clear-cut anyway, they argued successfully that "petri dishes" as large as 100 hectares should be included in an experimental manipulation.

With the assistance of Hans Matheve, Toon Spanhove and Luc Lens of the University of Ghent are studying avian demographic and nutritional responses to forest fragmentation within the Taita Hills of southeastern Kenya. Although the study is ongoing, they have graciously furnished me with an unpublished ptilochronology dataset for analysis and inclusion in this account. I have also benefited from access to an analysis of part of the dataset performed by Liesbeth Wiersma.

Once heavily forested, the Taita Hills are now marked by isolated woodland fragments surrounded by a landscape of agricultural fields and villages (Figure 4.12a). Within eight such fragments, ranging from 2 to 179 hectares, several species are being studied, one of which, the white-starred robin

Figure 4.12 (a) "Ngangao," a 92-hectare fragment of remaining indigenous woodland surrounded by fields and villages within the Taita Hills of southeastern Kenya, (b) personnel of the Ornithology Department of the National Museums of Kenya measuring birds in the field (c and d), respectively, adult and juvenile white-starred robins, a permanent-resident insectivorous species common within Taita Hills fragments. Photographs by Hans Matheve (a, b, d) and Toon Spanhove (c).

(Figure 4.12c, d), has been sampled in sufficient numbers for a preliminary analysis of growth bar width on original feathers.

Individuals of this permanent-resident insectivorous species were mist-netted, banded, measured, and had their two central original rectrices plucked and stored. Following the methods of earlier studies, if we restrict our attention to adult birds, we can assume the original feathers under consideration had been grown in the woodland fragment where they were collected. Growth bar width was calculated as the average of the means of 10 growth bars on the right and left rectrices, and tarsus length as the average of values for the left and right legs.

Across 219 birds in the eight fragments, multiple-regression analysis (Minitab 14.1) revealed a positive and significant correlation between growth bar width and tarsus length ($p = 0.02$). That is, as expected, larger birds tended to grow feathers at a greater rate per day. When the apparent effect of tarsus length was controlled statistically, a significant ($p = 0.04$) positive relationship emerged between growth bar width and fragment area, suggesting that robins living in larger fragments were in better nutritional condition (Figure 4.13).

Spanhove and Lens continue to collect records and envision a more extensive analysis to include information on habitat types and food supplies in the various fragments. Perhaps, inclusion of additional variables will help us understand, for example, why growth bar widths in the 12-hectare fragment, "Fururu," appear to be unusually narrow (Figure 4.13).

For the past decade I have been directing a project focused on the responses of forest organisms to the fragmentation and loss of woodlands in the agricultural landscape of north-central Ohio. Our study plot consists of a 15,400-hectare landscape dominated by row-crop agriculture (corn, wheat, and soybeans) in which occur about 100 woodlots of various sizes and two

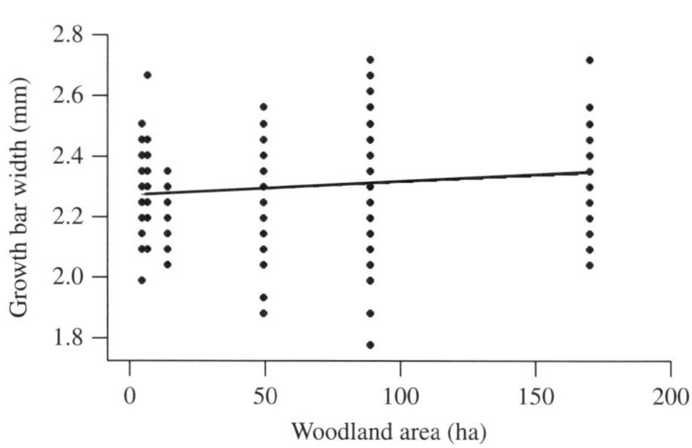

Figure 4.13 Growth bar width (controlled analytically for structural size of the bird) on the "original" central rectrices of adult white-starrred robins increases statistically significantly with woodland fragment size within the Taita Hills of eastern Kenya. Included is the fitted regression line. Calculated from an unpublished data set provided by Toon Spanhove and Luc Lens.

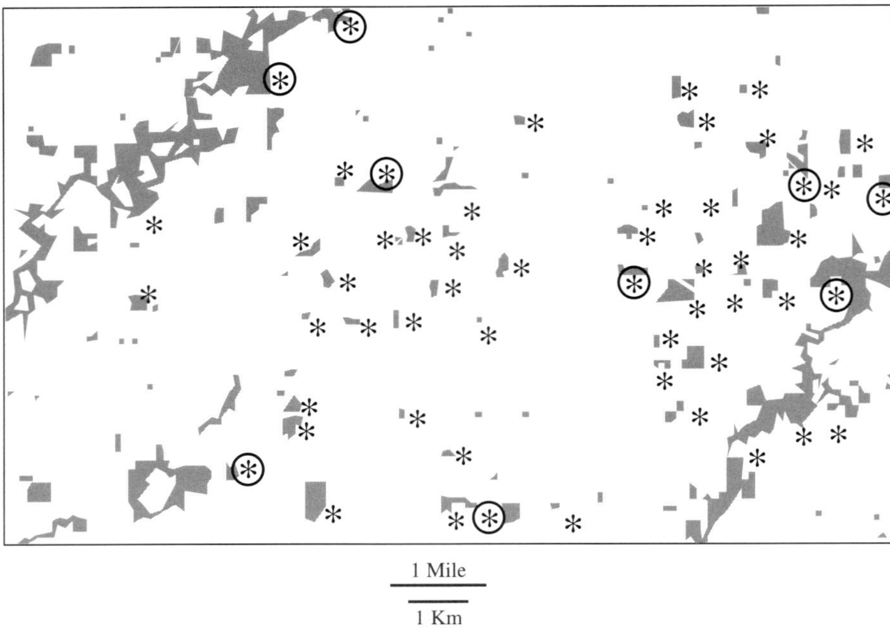

Figure 4.14 The Crawford County, Ohio, USA, study area, a 15,400-ha agricultural landscape, where ptilochronology was employed to index the nutritional condition of permanent-resident birds. The 9.6- × 16.0-km plot is flat and contains, within a matrix of row-crop agriculture, 101 woodlots ranging in size from 0.07 to 59.60 ha and two riparian corridors. Birds were banded in the 47 woodlot and seven riparian sites marked by asterisks. Food-supplemented sites are indicated by circles. From Doherty and Grubb (2002).

wooded corridors along rivers (Figure 4.14). Within these woodlots, Paul Doherty and I have been able to relate feather growth rates, as a measure of nutritional condition, to annual survivorship.

Each winter, we catch and band permanent-resident birds at 54 trapping sites, 47 in woodlots and seven along the river corridors. For four species, Carolina chickadee, tufted titmouse, white-breasted nuthatch, and downy woodpecker, we collected sufficient records over five winters to calculate annual survivorship (Doherty and Grubb 2002). For the chickadee, the nuthatch, and the woodpecker, the probability of survival was highest in large woodlots, low in small woodlots, and either intermediate or lower still in riparian woodlands. The presence of supplemental food at bird feeders maintained by some woodlot owners had a positive relationship with survivorship. For the tufted titmouse, most commonly foraging on the ground of the four species, survivorship was lower in years when heavy snow covered much of their food supply. It seemed reasonable to think that in these wintering birds, survivorship would be at least partially a function of nutritional condition, so we tested this idea by analyzing growth rates of feathers induced in various-sized woodlots, some with and some without supplementary food (Doherty and Grubb 2003).

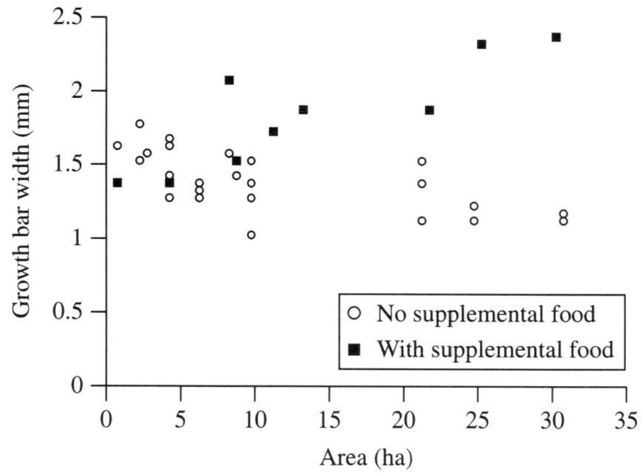

Figure 4.15 Daily growth of induced rectrices of Carolina chickadees with and without supplemental food in differing-sized woodlots. From Doherty and Grubb (2003).

We visited all sites in November and December to band birds and pull their two outermost original rectrices. (As mentioned earlier, the outermost rectrices of woodpeckers are vestigial, so we pulled the L5 and R5 from downy woodpeckers.) Then we re-trapped the sites in February–March to collect induced feathers. Growth bars were measured by student technicians with no knowledge of the feathers' provenance.

Feather growth rates partially matched the survivorship results. When controlled analytically for bird size and the day of winter when the original was plucked, growth bar widths of induced rectrices of chickadees were positively associated with the statistical interaction between woodlot area and food supply. That relationship was such that large, food-supplemented woodlots contained birds with the widest growth bars (Figure 4.15). We did find a negative relationship between presence of supplemental food and induced growth bar width in the nuthatch, but we can find no logical explanation for this result and think it might have been spurious. Also, as predicted, induced growth bar width was negatively correlated with snow cover in the tufted titmouse (Figure 4.16).

Presumably, birds that died of starvation during the course of the winter would have been those with quite narrow growth bars, in which case the distribution of daily feather growth rates available for us to study would have been truncated, biasing against finding significance of results.

In our previous analysis of annual survivorship, we had found that correlations between survivorship and woodlot size and presence of supplemental food were strongest in Carolina chickadees and weaker in white-breasted nuthatches and downy woodpeckers. The ptilochronology results are in line with such a distinction in survivorship. The chickadee's nutritional condition was correlated positively with the woodlot area × food interaction term such that large woodlots with feeders had the widest induced growth bars.

We found no positive effects of supplementary food in the titmice, nuthatches, or woodpeckers. All three of those species are larger than chickadees, and

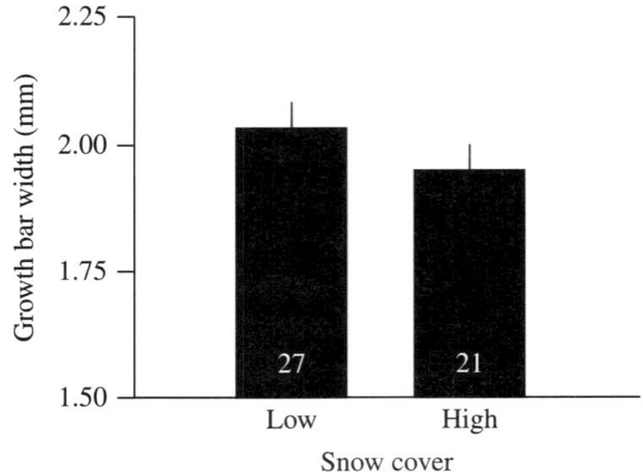

Figure 4.16 Daily growth (±SE) of induced rectrices of tufted titmice during winters of low (<5 days with ≥5 cm of snow on the ground at 0700 hours) and high (>25 days with ≥5 cm of snow on the ground at 0700) snow cover. Sample sizes are shown in the bars. From Doherty and Grubb (2003).

nuthatches and woodpeckers have distinctly different foraging niches focused on trunks and large branches of trees. The small size of the chickadee, its socially subordinate status in flocks, and its propensity to forage on small branches, sites with high thermoregulatory costs in comparison with those of other species, may have made supplemental food a more important factor for that species. The effect of supplementary food on feather growth was most clearly seen in large woodlots even though the greatest effect might be expected to occur in small woodlots. That result suggests that the addition of food was still insufficient to counteract the detrimental effects of small woodlot size.

Snow cover was significantly related to the nutritional condition of only the tufted titmouse. Tufted titmice spend a larger proportion of their time foraging on the ground than do the other three species we studied (Rybcynski 1977). Rogers and Smith (1993) suggest that ground foragers' food supplies are more affected by snow cover and, thus, in years of heavy snow, leaf litter is unavailable for foraging, contrary to years with little snow cover. Snow cover may have had particularly negative effects on titmice and not on the other species.

In summary, these results tie feather growth rate as a function of habitat quality directly to annual survivorship as a function of habitat quality. They are the clearest example reviewed so far demonstrating rate of feather growth to be a valid index of fitness.

We finish this review of ptilochronology and habitat quality with a pair of reports concerning possible effects of copper and lead pollution, respectively, on the nutritional condition of great tits. The first paper, on the relation between the yellow color of the breast feathers and feather growth rates (Eeva *et al.* 1998), is included here somewhat arbitrarily as it could have been assigned to the upcoming chapter on individual quality. However, the main thrust of the report concerns the effect of a pollutant on a presumably sexually selected trait, so I have elected to discuss the findings here in the context of habitat quality. Tapio Eeva and colleagues start from the premise that in

(a) (b) (c) (d)

Figure 4.17 (a and b) Views of pine forest in the vicinity of a copper smelter in Finland. (c) Female great tit on nest in a study nest box. (d) Nestling great tit very close to fledging. Photographs by Tapio Eeva.

many species, females prefer more colorful males as mates, therefore, male coloration is a sexually selected trait. They then note that the bright yellow coloration of the breast feathers in male great tits is due to incorporation of carotenoid pigments into the developing feather. Carotenoids cannot be synthesized by vertebrates, but must be assimilated from ingested food. Finally, they suggest that the principle sources of carotenoid pigments for great tits are the green caterpillars and sawfly larvae in their diet.

Eeva and colleagues studied the effect of sulfur dioxide and heavy metals emitted from a copper smelter in northern Finland on the coloration of great tits (Figure 4.17). They hypothesized that due to decreasing toxicity of pollutants with increased distance from the smelter, there would be positive correlations of distance with both density of caterpillars in the habitat and intensity of the yellow coloration of the tits. Second, if yellow coloration was a measure of nutritional condition indicating individual quality, birds with brighter yellow breast feathers should have wider growth bars on tail feathers.

The first part of the project focused on nestling tits and did not incorporate ptilochronology. Caterpillar and larval sawfly abundance was indexed by quantifying the rate of frass (i.e. defecation accumulation) from the insects under trees within tit home-ranges. Fifteen-day-old nestling tits were studied in nest boxes along air pollution gradients in three directions from the smelter (Eeva and Lehikoinen 1995). The nestlings were weighed and the intensity of the yellow color of breast plumage was determined using a six-category color scale with 1 = pale to 6 = bright.

The brightness of the yellow coloration of nestling breast feathers and the density of caterpillars both increased with distance from the smelter. These correlations supported the link between carotenoids in caterpillars and nestling color. Furthermore, body mass of these 15-day-old nestlings also increased with distance from the point source of pollutants, buttressing the link between amount of carotenoid-bearing insects ingested and color of the nestlings.

The second part of the project in the same locations employed ptilochronology to investigate whether there was a link between nutritional condition and the breast color of the plumage grown during the normal molt of body feathers that occurs some weeks after fledging. Juveniles were caught in the autumn, their breast coloration determined, and the average width of nine growth bars on the right fifth rectrix calculated. Thus, the analysis was confined to original flight feathers.

The relation between breast color and growth bar width approached but did not reach statistical significance for either male or female juveniles (P values of 0.13 and 0.19, respectively). The authors conclude, therefore, that breast coloration did not indicate nutritional condition in juveniles of the species during the time of the molt. Upon reflection, though, it seems quite possible that there was a link between feather growth and color of the breast plumage. Eeva and colleagues used the average of nine growth bars to calculate daily rate. The total length of a feather, however, is a cumulative function of many days of growth, perhaps, about 35 days in the case of these juvenile great tits. Thus, the nonsignificant relationship between daily growth and feather color, when compounded by the total number of days of growth reflected in total feather length, led to significant positive correlations between the total length index of nutritional condition and plumage brightness for both males and females, $P = 0.04$ and 0.03, respectively. We will return to the issue of nutritional condition and plumage color in great tits later in another study in Chapter 7, a study that found a significant relationship between growth bar width and brightness of the yellow breast plumage.

The second "pollution" study took a controlled laboratory approach to the hypothetical relationship between lead pollution and avian nutritional condition (Talloen *et al.* 2005). Willem Talloen and colleagues housed Belgian great tits individually in cages so they were visually, but not acoustically, isolated from each other (Figure 4.18a). Birds were randomly assigned to either an experimental group given water containing 10 parts per million lead or to a control group given water without the additive. Nine days after the experimental and control treatments had begun, the second-to-outermost right and left rectrices (L5 and R5) were pulled from each bird.

The study's objective was to assess both feather growth rate and extent of bilateral symmetry as a function of two factors, (1) the presence of the lead pollutant in the drinking water, and (2) whether the two induced feathers grew in at the same time. To accomplish the latter objective, birds in both experimental and control groups were randomly assigned in equal numbers to

(a)

(b)

Figure 4.18 Effects of lead intake on growth rate and bilateral symmetry of induced rectrices in the great tit. (a) Laboratory set-up showing holding cages with bottles dispensing either water containing 10 parts per million lead or water alone. (b) Technique for measuring the length of a rectrix daily as it grows on the bird. Photographs by Willem Talloen.

one of the following three subgroups, (1) left feather pulled seven days before the right one, (2) right feather pulled seven days before the left, and (3) both left and right feathers pulled at the same time. Thus, the experiment had six subgroups, three each of both lead-treated and control birds.

Daily feather growth was calculated as the average width of five consecutive growth bars, and total feather length was also measured (Figure 4.18b). For each bird, daily growth rate and total length were calculated as the average of the values for the left and right feathers. Fluctuating asymmetry was taken as the logarithm of the absolute difference in growth bar width or feather length between the L5 and R5 rectrices.

This laboratory study furnished the opportunity to assess two variables not accessible in field studies with free-ranging birds. First, the birds were examined in the hand every day until the induced feather emerged, a procedure that furnished an index of the lag time to growth of an induced feather. The time lag measured was the interval from the time of plucking the original feather until emergence of the induced feather from the surface of the skin. Start of growth in the follicle, several millimeters below the skin surface, would not have been accessible to external monitoring. Second, a feather was defined as fully grown after it had had the same total length, on the bird, for four days. This second measure was particularly important in determining whether bilateral asymmetry was related to the time lag in induction of L5 and R5 rectrices.

In the birds from which the left and right rectrices were pulled at the same time, the lead-exposed group delayed the re-growth of the induced feathers by about four days, a significant difference from the control group. In the asynchronously plucked subgroups, there was no difference in lag time to emergence for the two induced rectrices between the experimental and control group.

The daily growth of the induced feathers measured both on the bird and as the average width of growth bars was significantly less in the lead-treated group. Growth bars were about 0.2 mm narrower in the lead-exposed group (Figure 4.19a). By contrast, the level of fluctuating asymmetry of full-grown feathers was not affected by either asynchrony of pulling the left and right original rectrices, or exposure to lead pollution. If anything, there was a trend for less fluctuating asymmetry in the lead-exposed group (Figure 4.19b).

Several important points arise from this carefully conceived study. First, there were larger differences in growth bar width and total feather length

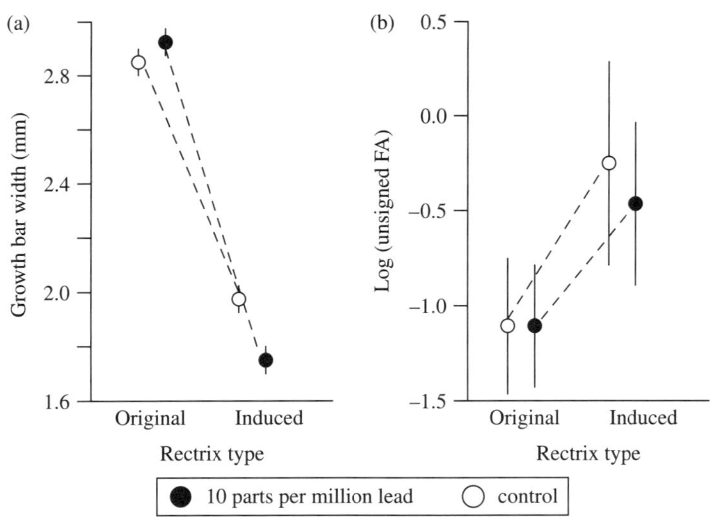

Figure 4.19 Means ± SE of (a) rectrix growth bar width, and (b) fluctuating asymmetry (FA) of rectrix total length in experimental great tits with drinking water containing 10 parts per million lead and in control tits with unpolluted drinking water. Original rectrices were grown in the wild before the birds were captured, while induced rectrices were grown during the experiment. From Talloen *et al.* (2005).

between original and induced feathers than between induced feathers of control and lead-exposed groups. Talloen and his colleagues argue that growth of original and induced feathers involve different processes, on the one hand normal development and on the other a repair process that is likely to involve different biochemical and physiological mechanisms. Also, as we have seen, an influence of hormonal levels probably plays a role.

The second point is that, at least for this feather type and this form of pollution, fluctuating asymmetry was much less sensitive to environmental stress than was induced anabolism in the form of feather growth.

Third, the induced feather became noticeable beyond the level of the skin at a minimum of 10 days after the original feather had been plucked. We will see in an upcoming project on nestling growth in tree swallows (McCarty and Winkler 1999) that reduced food intake appears to be able to slow feather growth as much as four days later. Thus, as Murphy and King (1991) pointed out, there does seem to be some delayed effect of a given level of nutrition on feather growth. In this tit study, birds had been assigned to treatment groups at least 10 days before the first sign of induced feathers. Thus, it seems likely that the birds had been exposed to treatment or control for an adequate period to eliminate any effect of pre-treatment condition on feather growth. Nevertheless, it would be quite useful to know the lag time between pulling the original feather and the first sign of induced feather growth and extension at the level of the follicle some millimeters below the skin of a bird.

Talloen and his co-workers concluded that width of feather growth bars seems a particularly useful biomonitoring tool, at least for great tits, because (1) it can be measured on fully grown feathers, (2) it can be measured with high reliability, and (3) growth bar width is known to provide a reliable measure of daily feather growth. In their study, the variance in daily growth measured on the bird that was explained by width of growth bars was 99.6%.

4.5 *Summary*

Results of these nine studies of habitat quality tell us that rate of feather growth appears to be a sensitive measure of nutritional condition. In two studies (Brown *et al.* 2002; Doherty and Grubb 2003), feather growth rate has been related directly to the survivorship component of fitness. It is important to point out, however, that most of these studies fall short of the early suggestions for proper use of the method (Grubb 1989, 1992). Six of the nine studies relied solely on original feathers, and one study lumped original and induced feathers together for analysis. In most projects, and in all of the field projects except that with scrub-jays, exactly when and where an original feather had been grown were unknown. None of the field projects used random assignment of birds to treatment and control groups, not surprising in a study of habitat quality. These various shortcomings bring a bit of uncertainty to the results even though, as a group, these projects indicate that feather growth rates can index quality, particularly relative quality, of avian habitats.

5 *Nutritional consequences of self-cached food*

In the temperate and boreal zones, animals from honey bees to beavers store food during the growing season for use during the winter when food is non-renewing. Birds are among the better-studied examples of such food-hoarders. During the last several decades, research has focused on two aspects of caching. The first line of work has searched for the mechanisms behind some birds' apparently extraordinary ability to remember the location of and retrieve hundreds or even thousands of individual caches they have previously sequestered within their homerange. (Russell Balda, a pioneer of this line of research, once determined that Clark's nutcrackers, his study animal, could remember where they had cached food considerably better than could his graduate students!) The second line of investigation has focused on the adaptive value, the fitness benefits, of caching and it is to this second line that ptilochronology has made a contribution. The four studies detailed below all demonstrate, through increased rate of induced feather growth, that birds derive a nutritional benefit from retrieving food items they have previously cached.

Tom Waite was the first to realize the utility of ptilochronology for studying caching (Waite 1990) and incorporated the method into a larger study of foraging behavior of Alaskan gray jays. During summer and autumn, when food is abundant, jays store thousands of food items, everything from berries to bits of dead moose, in separate locations within their year-round territories. They then retrieve and eat these cached items during the winter when far-northern populations, such as the one that Waite studied at 66° North Latitude, have only a few hours of dusk-like conditions each day in which to forage before darkness again descends.

Waite predicted that jays given supplemental food to cache would retrieve such caches during the following winter, growing wider growth bars on induced feathers in the process. Thus, feather growth rate should indicate the better nutritional condition of such food-supplemented jays.

Waite randomly assigned territorial groups of jays to either treatment or control groups. (This was the first ptilochronology field study to employ the crucially important random assignment of birds to treatment and control groups.) In October, he made pelleted dog food available to the birds in the treatment group, food that they cached assiduously. The jays defend group territories and about 680 pellets were provided per jay in each "fed" territory. Feeders in the territories of the control group were left empty, although such birds were baited with the small number of raisins necessary to trap them.

In mid-November, Waite stopped the supplementation program and captured and plucked the R6 rectrix of both experimental and control jays. Then in early February, he re-baited the feeder-traps and gathered induced feathers from eight jays in four social units at control sites and from 13 jays in seven social units at experimental sites. Because birds in the same social unit on the same territory should not be considered statistically independent, feather growth values of all birds in the same social unit were averaged for analysis. Adults grow their original rectrices during the fall molt while juveniles begin growing theirs while still in the nest. In order to control for this difference in time when original feathers are grown, Waite used in the analysis only feathers of adults.

There was no statistical difference between control and treatment groups in growth bar width, feather length, or feather mass of original rectrices. By contrast, growth bar widths showed that adult jays in supplemented territories grew induced feathers 9.2% faster (Figure 5.1). The induced rectrices of supplemented adults also grew 7.1% longer and 10.6% heavier (Figure 5.2).

So far as I know, Waite's was the first study marked by a complete lack of growth of an induced feather. While every bird in six cache-supplemented social units grew an induced feather, in three of the four cache-unsupplemented groups at least one (adult) jay failed to regenerate the feather. The complete failure to regenerate a feather gives us a clue that in times of severe nutritional stress, birds may have another option besides regenerating a feather slowly, they may conserve the maximum amount of energy by not growing it at all. It would be interesting to know whether these non-growers started up regeneration with the more benign conditions applying later in the spring. In

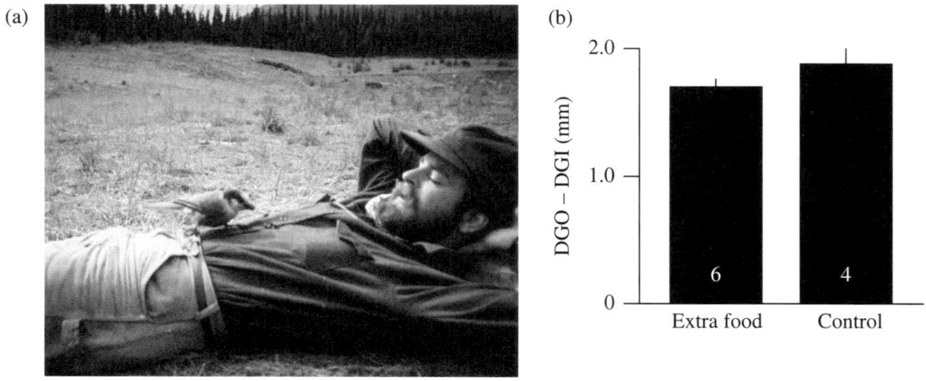

Figure 5.1 (a) Tom Waite with Alaskan gray jay collecting raisins to cache in its territory. (b) Mean ± SE difference in growth bar width between original and induced R6 rectrices of gray jays that had previously cached extra food in their territory and those that had not. Sample sizes are shown in the bars. The induced rectrices of birds with access to extra food were also significantly longer and heavier. (a) Photograph by the author. (b) from Waite (1990).

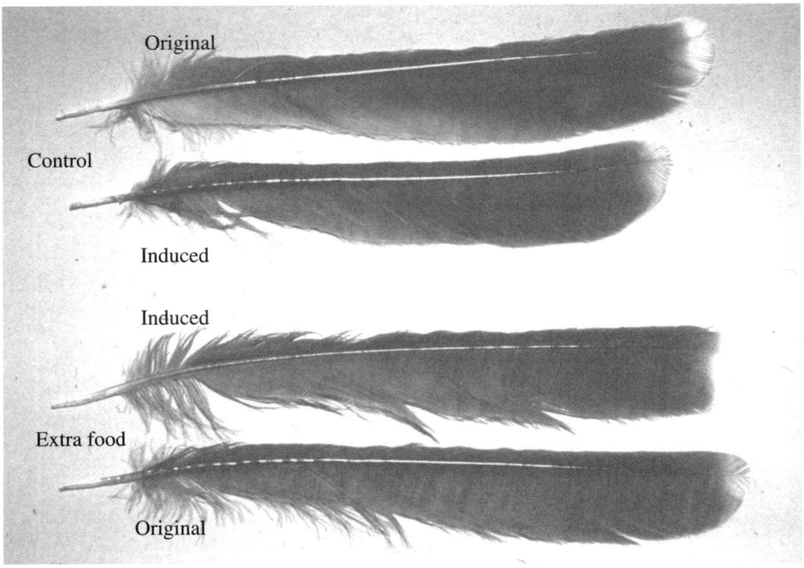

Figure 5.2 Examples of original and induced R6 rectrices of Alaskan gray jays that had (Extra food) or had not (Control) been given extra food to cache in their territory during late Autumn. While the original rectrices are about the same size with similar growth bar widths, the induced feather from the food-supplemented bird is longer and has wider growth bars. Photograph by Thomas A. Waite.

any case, in a beautifully designed experiment carried out under extremely rigorous conditions, Waite provided us with very strong evidence for the fitness benefits to birds of caching food.

Jan-Åke Nilsson and his colleagues in Sweden used ptilochronology to extend our knowledge about the implications of caching for fitness (Nilsson *et al.* 1993). Applying their work to Eurasian nuthatches, they wanted to establish the existence of long-term hoarding in a small songbird. Previously, such long-term hoarding and retrieving were best known in the larger jays and crows. The Swedish workers were also curious to see if the nuthatches might have some strategy for using stored food in winter only under the more difficult circumstances of very cold, windy weather. Finally, they used ptilochonology to assess the possible effects of hoarding on the nuthatches' nutritional condition under the assumption that animals in better nutritional condition might be headed for higher lifetime reproductive success.

Like their ecological equivalent in North America, the white-breasted nuthatch, Eurasian nuthatches spend the year on mated-pair territories where, in autumn, they regularly hoard natural seeds in large quantities. Hoards are made individually, often in the deeply fissured bark of large deciduous trees. The birds particularly key in on beech mast, but also use hazel nuts. Nilsson *et al.* were perhaps fortunate that during the winter of their caching study, the beech mast crop in southern Sweden was very poor so that any nutritional effect of the supplementary food they provided would be more detectable

In the autumn, they assigned five woodlots containing 16 pairs of nuthatches to the treatment group and six woodlots with 14 nuthatch pairs to the control group. (We are not told explicitly whether treatment and control groups were randomly assigned.) On the first day of November, feeders filled with sunflower seeds were placed in the treatment territories and kept filled for 30 days as the nuthatches removed and cached seeds. About 18 kg of seeds disappeared from each feeder during this period. No supplemental food was provided in the control woodlots.

During the interval from 10 to 84 days after stopping the provisioning, the researchers monitored the rate at which focal birds in each woodlot retrieved stored seeds, confining their attention to females whose behavior was not confounded by territorial displays as was the males'. For each seed retrieved, they recorded whether the female ate it or re-cached it in another location. Air temperature was recorded at the start of each observation session.

Between mid-January and the first of February, well after they had emptied the feeders in the treatment woodlots, the researchers caught 34 of the nuthatches, banded them, and pulled the left and right fourth rectrices. Then, between early March and early April, they re-caught all 34 birds to retrieve the full-grown rectrices. All feathers were coded and measured in a "blind" fashion, and analyses employed the mean value of the two feathers pulled from the same bird.

Over the course of the winter, the supplemented females that were watched retrieved and ate more sunflower seeds in lower temperatures. This relationship leads to the conclusion that the birds refrained from using their stored food resources during more benign weather, husbanding them for cold periods. They might have done this in any of three ways. First, they could have noticed seeds by chance in all sorts of weather, but not taken them out of their storage places when it was warmer. Second, they could have found seeds by chance in all kinds of weather, but searched for them harder in cold weather. Third, they could have known the locations of all the seeds all the time and just refrained from retrieving them in warmer weather. It would be interesting to know which alternative or alternatives was or were correct. Incidentally, this prudent use of caches is a clear benefit of being territorial. No other nuthatch except the mate is around to eat seeds that are being saved for when they might really be needed.

Growth bar width, feather length, and feather mass of original feathers were all statistically identical between control and treatment nuthatches. However, all of these measures of induced feathers were greater in the supplemented birds (Figure 5.3). As Figure 5.3 illustrates, the difference between control and treatment groups was greater for mass and length than for daily growth, from which we might conclude that supplemented birds put more material per day into their induced feathers and grew them for more days. This Swedish project adds to the story by suggesting that not only does

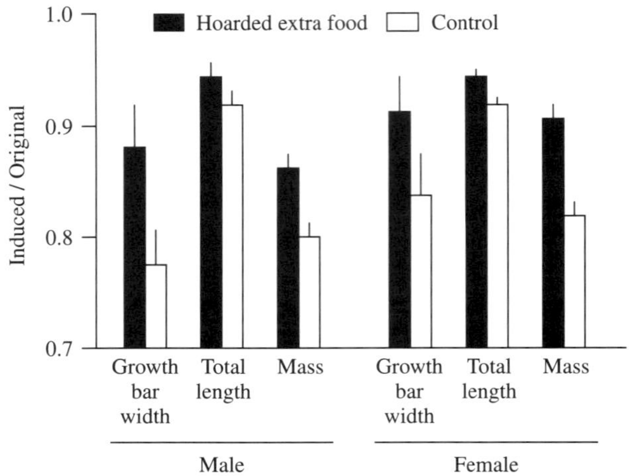

Figure 5.3 The ratio of induced to original rectrix growth for male and female Eurasian nuthatches. Birds in one group were allowed to cache sunflower seeds in their territories early in winter. Retrieving these hoarded seeds improved their nutritional condition over the course of the next several weeks compared with birds in the control group. All six differences between treatment and control groups were statistically significant. From Nilsson *et al.* (1993).

caching increase nutritional condition, but that birds can actually adjust their nutritional condition over a number of weeks by regulating their rate of cache retrieval as a function of ambient temperature.

In Part I of this account, I mentioned that Anders Brodin used radioactive food to demonstrate conclusively that one growth bar corresponds to 24-hours' growth of a feather. Brodin and Jan Ekman went on to apply this technique in a study of caching and cache retrieval in Swedish willow tits (Brodin and Ekman 1994). The question of interest was the following: do other birds in a social group derive a benefit from retrieving and eating food cached by any given member of the group? That is, does cache pilferage occur? The first step in finding out was to band and pull a tail feather from all the birds (usually four) in territorial groups of this permanent-resident species. The researchers then allowed one particular bird in each group to cache 20 seeds labeled with the amino acid, ^{35}S-containing cysteine. The other members of each flock were allowed to cache only unlabeled seeds. Two months or more after they had been plucked, all birds were recaptured and the induced rectrix plucked. Brodin and Ekman predicted that if only the cacher of seeds derives the benefit of retrieving and eating them at a later time, autoradiographs should reveal radioactive growth bars (Figure 1.3) on the induced rectrix of only the cacher of labeled seeds.

According to the records of radioactive growth bars, during the six to 40 days after the caching event, the cacher of radioactive seeds retrieved such seeds on an average of 5.1 separate days (N=9 flocks). By contrast, each other

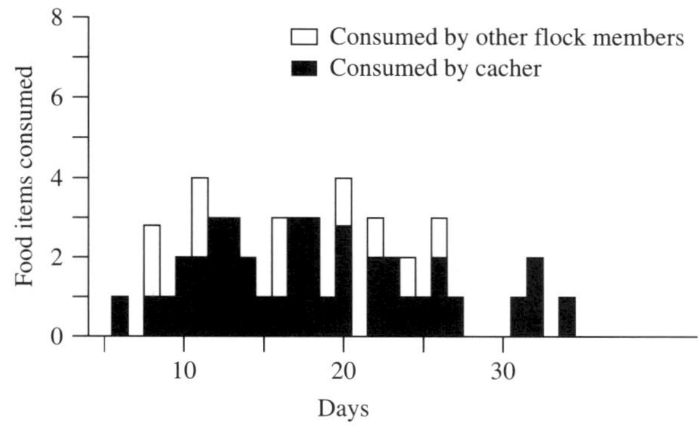

Figure 5.4 As indicated by radioactivity in an induced rectrix, consumption of radio-labeled sunflower seeds by the one willow tit in a flock that had cached such labeled seeds, and by all other flockmates. Over the course of the study period, the cacher retrieved about four times as many of its cached seeds as did all of its flockmates combined. From Brodin and Ekman (1994).

member of a social group, a bird that had not cached labeled seeds, retrieved a labeled seed on only 1.0 day during the 34-day period (Figure 5.4). As Brodin and Ekman point out, because there was no lessening of the advantage to the cacher by the end of the monitoring period, it seems likely that a cacher enjoyed almost exclusive access to its own caches throughout the course of the winter.

It is well known that territorial flocks of willow tits do not contain close blood relatives. Therefore, while speculating about the behavioral mechanism or mechanisms allowing the cacher to find its caches, the researchers concluded that their results indicate that the fitness benefit of caching is selfish. The non-related members of a group do not benefit to any appreciable extent from each other's caching activity. Before leaving this study, we should complement the authors for using the flock rather than the individual as the primary sampling unit for statistical analyses.

The most recent ptilochronological chapter related to the fitness advantages of caching comes from a study by Jan Ekman and students (Ekman *et al.* 1996) of Swedish Siberian jays wintering at 65° North Latitude. Unlike willow tits, the jays live in year-round social groups, groups that often contain relatives. Specifically, wintering groups contain a mated pair of adults that bred at the same site the previous summer and, usually, one or more of their offspring from that summer. Often, also, winter social groups contain one or more non-related juvenile jays that have dispersed into the area from elsewhere and taken up residence with the group of related birds.

Similarly to Tom Waite's gray jays in Alaska, Siberian jays cache great numbers of food items such as berries and seeds. Ekman and colleagues wanted to know if the jays employed a strategy of gaining indirect fitness benefits. That is, did the birds behave so as to further the lifetime reproductive success of relatives, the indirect component of fitness, as well as their own lifetime reproductive success, the direct component of fitness? The researchers used Brodin's field methods and radioactivity analysis to investigate the matter, feeding small pieces of ^{35}S-labeled suet to one juvenile and unlabeled food to all the other birds in each social group. All birds cached such food items extensively. While groups containing various numbers of relatives and non-relatives were studied, we will concentrate on flocks that contained a mated adult pair, at least one offspring and at least one unrelated juvenile. If cachers derive an indirect fitness benefit by having their relatives retrieve and consume their food caches, thereby lessening risk of starvation for the relatives during the winter, then first-order relatives (parents and siblings) of the cacher of radioactive seeds should show more radioactive growth bars than non-related members of the same social group.

First, results agreed with the willow tit study in showing that the cacher had a clear advantage in retrieving its own caches, in this case a seven to one advantage (Figure 5.5.). Figure 5.5 indicates that the advantage to the cacher was likely to continue after the period that could be monitored by feather growth. (Such a limitation indicates what could be done if feathers were plucked from the same bird at overlapping time intervals to extend the monitoring period.) Second, among the non-hoarders, first-order relatives of the hoarder showed no increase in cache retrieval over non-relatives. Eleven first-order relatives retrieved only one labeled food item, which worked out to 0.09 radioactive growth bars per individual in six groups, a rate not statistically different from the retrieval rate of 15 non-relatives, an average of 0.40 radioactive growth bars per individual in seven groups.

Siberian and gray jays both use a hoarding technique that might make the favoring of relatives difficult. Birds cache independently of one another throughout the group territory, and often conceal the caches in foliage. Such caches would seem to be as difficult for relatives to locate as for non-relatives.

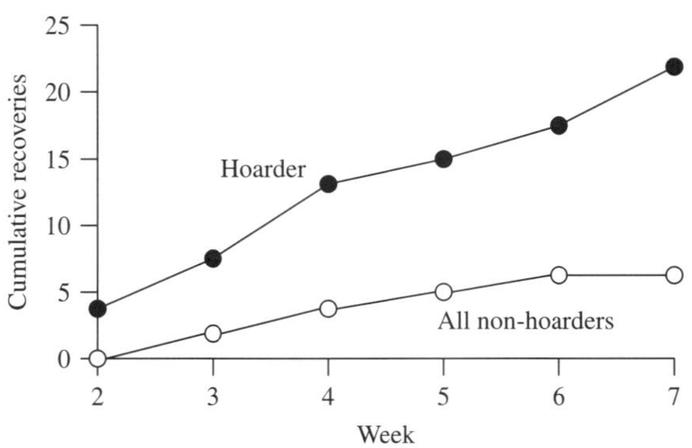

Figure 5.5 As indicated by radioactivity in induced rectrices, cumulative recoveries of hoarded food items by the Siberian jay that hoarded those items and by all other members of its territorial group. From Ekman *et al.* (1996).

5.1 *Summary*

This series of caching studies brought well-designed field experiments to ptilochronology. Random assignment of control and treatment groups, use of induced feathers and careful attention to plucking dates are all design features that reduced ambiguity of results. The series of papers also illustrates how scientific knowledge accumulates from steps made by a community of workers. In this case, we started with the finding that cachers benefit nutritionally from their caching, we then learned that cachers may manage their caches to be used only under the more trying environmental conditions they experience and that among non-relatives living in a social group, each bird benefits selfishly from its own caching activity, and finally, we discovered that indirect fitness benefits of caching do not seem to occur, at least in the one species that has been examined to date. Relatives of the cacher retrieve no more of its caches than do non-relatives.

6 *Social behavior*

6.1 *Intraspecific social behavior*

Birds living together in a social group have been studied with an eye to determining who is enjoying a fitness benefit and who might be suffering a fitness cost from such sociality. Ptilochronology has played a role in determining these benefits and costs. It turns out that most such studies have been carried out during the non-breeding season. Many avian species found in flocks when not breeding break into territorial pairs during the breeding season, and in subsequent chapters I will discuss nutritional condition related to social aspects of reproduction.

So far, the use of ptilochronology in the study of social behavior during the non-breeding season has involved a paucity of species. Most of the work to be covered here deals with titmice, nuthatches, and sparrows. There is a good reason for this, namely the high degree of overlap between the winter ranges of these species and the locations of academic researchers in Europe and North America. Particularly in the realm of ptilochronology, we badly need comparative records from other regions of the world. With that caveat in mind, I will start by examining some aspects of social behavior within a species, then turn to studies of inter specific social behavior.

Olav Hogstad of the University of Trondheim has applied ptilochronology in several inventive ways to non-breeding Norwegian willow tits. To preface accounts of his work, I will review the portion of willow tit life history concerning fledging, dispersal, and establishment in winter social groups. Much of this material applies to other species of titmice and chickadees as well.

Willow tits leave their natal territories within several weeks of fledging. They then move about the landscape, apparently searching for a flock in which to insert themselves for the winter. In Norway, a full complement of birds in such a flock is four to six, arranged in dominance-ordered male-female pairs. Apparently, a dispersing juvenile is looking for a flock without the full complement of members of its sex. When it finds such a flock, it joins it and assumes a particular position in the dominance hierarchy. As you can imagine, it is very difficult to study the ephemeral events associated with leaving the natal territory and joining a flock elsewhere. They occur in a brief time span and are accomplished by highly mobile animals. Hogstad has shown, however, that the largest, heaviest fledgling of a brood tends to disperse from the natal territory first, and that juveniles that fledge early are more likely to find winter-flock openings, a sort of first-come, first-served syndrome. As there are far more fledglings produced than there are openings in winter flocks, it appears that early fledging and dispersal are favored by natural selection.

Hogstad's (1990) first question addressed by ptilochronology asked whether young willow tits that attained higher social dominance positions in a flock had been in better condition as late nestlings and fledglings than those that attained lower social dominance in flocks. (Similar to the Florida scrub-jays we examined earlier, willow tits begin growing their tail feathers while still in the nest and finish growing them some weeks after fledging.)

Dominance status is often ascertained by watching and tallying interactions between color-banded birds. A bird that is chased by another, is supplanted by another from a feeding site, or waits until another bird leaves a feeding site before feeding there itself is considered to be subordinate to the other bird.

In September, Hogstad plucked the R6 original rectrix from juveniles that had been members of a social group for about two months (Figure 6.1). He examined dominance position (in descending order: juvenile male 1, juvenile male 2, juvenile female 1, juvenile female 2) as it related to growth bar width of the original feather and found that the two variables were not correlated statistically. Thus, he concluded that the nutritional condition of a willow tit in its early life is at best of only minor importance in determining its dominance rank in a resident flock. As we know from other work with the closely related marsh tit (Nilsson 1989), the date when a young bird disperses from its natal site and begins looking for a flock with an opening is very important for determining flock-joining date and subsequent dominance rank in a flock. Although very difficult to obtain, it would have been useful to have the dates when all of Hogstad's birds dispersed from home.

Black-capped chickadees in North America have essentially the same pattern of juvenile dispersal, winter-flocking, and territorial breeding pairs as

Figure 6.1 (a) A willow tit. (b) Olav Hogstad weighing a willow tit during his studies of the species in central Norway. Photographs by Olav Hogstad.

willow tits in Eurasia. Harry van Ort and Ken Otter (2005) recently used ptilochronology to study nutritional condition of young dispersing chickadees in British Columbia, not in relation to intra-flock dominance status, but in relation to quality of the habitat in which they settle. The researchers tested the notion that the young birds followed a "despotic distribution" (Fretwell and Lucas 1970; Fretwell 1972) in where they chose to spend their first winter. Despotic distribution theory holds that the distribution of animals in space is determined by the outcome of dominance encounters, and would predict that chickadees in better condition as dependent juveniles in their natal areas would, upon dispersing, use that better condition in a despotic way to secure better wintering locations. Birds in relatively poor condition would then be relegated to worse habitat. Van Ort and Otter monitored chickadees in two types of habitat, a recently clear-cut "disturbed" site and a mature "undisturbed" woodland. They knew from their work during previous breeding seasons that reproductive success was lower in the disturbed site, primarily due to increased nest desertion there caused by an inadequate food supply. They predicted that if dispersing chickadees arranged themselves in a despotic fashion, birds in better condition at dispersal, indicated by wider growth bars on rectrices grown as nestlings and dependent juveniles, would be more likely to be found in the better, undisturbed, habitat.

Female chickadees agreed with despotic distribution theory. After any possible effect of body size had been controlled by including tarsus length in the analysis, females in the undisturbed site possessed significantly wider growth bars on tail feathers than did those in the disturbed site (Figure 6.2). Thus, it appears that young females competed for better habitat in which to settle during the autumn dispersal period, those in better nutritional condition winning out.

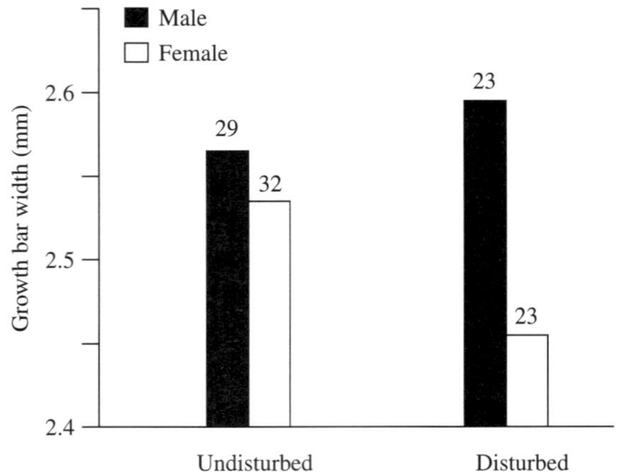

Figure 6.2 Mean growth bar width on original rectrices of first-year male and female black-capped chickadees wintering in recently clear-cut (disturbed) and mature (undisturbed) forest. The difference in growth bar width between the two habitat types was significant for females, but not for males. Sample sizes are shown above the bars. From van Ort and Otter (2005).

By contrast, the young male chickadees did not follow the theory; growth bar widths were statistically the same in males found in the two habitat types.

It is well known in chickadees (Weise and Meyer 1979) and many other songbirds that natal dispersal distance is greater in females than in males. Van Ort and Otter suggest that, compared to that of females, the decision to settle in males is less influenced by habitat quality and more by available openings in winter flocks. A male that finds a flock opening is less likely to leave that opening to search for higher quality habitat. Such a difference would account for both the smaller dispersal distance of all males, on the average, and the less discriminating habitat selection by high-quality males.

"Floater" has been a name given to individual birds of social species that have not inserted themselves into the dominance structure of a flock. Instead, they spend the winter "floating" from flock to flock, and are alone for much of the time. Such birds have often been late in fledging and have found no openings in flock structures. In a companion project to that discussed above, Hogstad (2003) asked whether willow tit floaters might exist under a different nutritional regime than members of flocks. They would not have to compete socially in a flock for feeding sites and so might be under better nutritional condition. On the other hand, when alone they would not be able to share the task of being vigilant for predators, as occurs in a social group. Because the necessary increase in time spent being vigilant themselves would reduce the time they had to look for food, they might be in poorer nutritional condition than members of their species living in flocks. Hogstad measured growth bars on both floaters and subordinate (juvenile) members of flocks to test the idea that floaters were in poor nutritional condition compared to flock members.

The outermost left tail feather of floaters and of the most socially subordinate male and female within flocks were collected from late September into October. Six to eight weeks later, Hogstad recaptured a number of birds and removed the induced rectrix from each. He mentioned that in his study area in the mountains of central Norway, winter conditions set in during mid-October, so the induced feathers grew under conditions of snow cover and periodic spells of very cold weather. As birds might produce different size tail feathers for no other reason than that the birds were of different size structurally, Hogstad attempted to remove variation due to size by dividing induced growth bar widths by growth bar widths on original feathers from the same follicles. (See discussion in Chapter 7 about the consequences of such a standardization technique.)

There was no difference in the amount of fat carried by floaters and by the most subordinate male and female flock members, but there was a clear difference in growth bar widths of induced feathers. Floaters had reduced rates of feather growth (narrower growth bars) compared to the most subordinate flock member of the same sex (Figure 6.3). Floater males and females, respectively, grew their feathers at a 12% and 9% slower rate than did the subordinate birds in flocks.

The question of why birds join winter foraging flocks has long been of interest (e.g. Ekman 1979). It has been known that in species such as willow tits, the

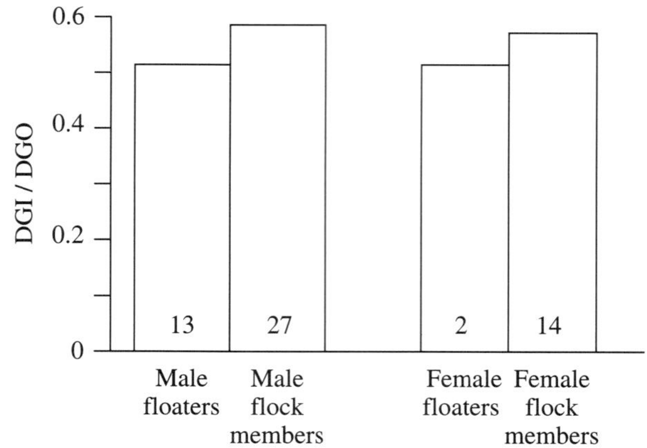

Figure 6.3 Ratio of growth bar widths of induced (DGI) and original (DGO) left outer rectrices (L6) of willow tits that were "floating" outside of social flocks or were the most subordinate flock member of the same sex. For both sexes, the difference in feather growth of floaters and flock members was statistically significant. Sample sizes are shown in the bars. From Hogstad (2003).

dominant pair in a winter flock secures the same area as a breeding site the following spring after having expelled the flock subordinates from the area. It has also been known that if a dominant bird dies during the winter, the next bird of the same sex down the dominance hierarchy moves up to take its place, in the process inheriting the dead bird's mate and breeding opportunity. So, the potential for breeding at the site clearly makes it adaptive for some juveniles to join a flock in hopes of moving up the dominance ladder during the winter. Such a mechanism may also provide the answer for why flocks are limited to 4 or 6. The third-ranked individual of each sex gets breeding rights only if the two birds higher in the hierarchy die during the winter. If we assume that the winter mortality rate is about 50% in these birds, a reasonable estimate (Ekman 1984), then the probability of a third-ranked bird of one sex in a flock of six birds inheriting the breeding site is $0.5 \times 0.5 = 0.25$. A fourth-ranked bird would have one chance in eight of doing so. Natural selection may have selected floaters to continually look for new openings in a number of flocks rather than staying put as the number four bird, say, in any given flock. What we had not known before Hogstad's work was whether low-ranked birds in flocks were paying a nutritional price for being there instead of adopting the floater life-style. Now we know that not only are their chances of securing a territory presumably better if they are in a flock, so is their nutritional condition during the winter.

It is no wonder that natural selection has willow tits and other titmice and chickadees starting reproduction early in the year, long before migrants begin breeding. The early-breeding bird has offspring that, after fledging, disperse early and are therefore more likely to find a position in the next winter's flock. Flock membership, in turn, gives them better nutritional condition through the winter and a chance to breed at the site the following spring.

The last part of Hogstad's investigation of nutritional condition in willow tits concerned the adaptiveness of the male-female pair-bond through the winter. In titmice and chickadees, the members of the alpha or most dominant pair, in particular, usually remain close to each other within a flock. Hogstad (1992) sought an adaptive reason for such social cohesiveness months before the breeding season would begin. From his own work and the work of others, he knew that within a flock of four to six birds arranged into two or three male-female pairs, all males socially dominated all females and adults socially dominated all juveniles of the same sex. Hogstad suspected that by having the dominant male in close attendance, the dominant female might gain protection from aggression by the subordinate males, and that such protection might increase her nutritional condition. Such a benefit to the alpha female would also benefit her dominant mate if it allowed her to begin egg production earlier in the spring. (We have seen how important early breeding leading to early fledging is in this and other titmouse species.)

Hogstad divided his observations of an adult female into cases where her dominant mate was either less than or greater than 5 meters from her in the forest. He predicted that when the male was close to her, she would be subject to fewer attacks from subordinate males than would the other females in the flock, and she would be less vigilant for the aggression of the subordinate males and thus spend more time foraging. Because of such effects of the adult male's presence, the nutritional condition of the adult female should be better than that of any of the juvenile members of the flock, male or female.

Hogstad studied dominance of color-banded birds in the usual way by recording chases, supplanting attacks, and instances of waiting at a feeder for another bird to finish feeding. As predicted, when her mate was nearby, the adult female was involved in few dominance interactions. She was chased or supplanted by the juvenile males less often than was either juvenile female. When her dominant mate was near, the percentage of the total time the adult female spent foraging rather than being vigilant increased by 12% over when he was not near. Presumably, the adult female was reducing her vigilance for approaching juvenile males.

Hogstad collected original feathers early in October and re-caught the birds in mid-January to pluck the induced replacements. As we have seen done before, he used the ratio of growth bar widths on induced and original L6 rectrices to standardize for differences in bird size.

Correlated with his top position in the dominance hierarchy, the adult male showed the greatest value of the growth bar index. However, even though the alpha female was number four in the hierarchy (behind the dominant male and both juvenile males), the mean ratio of growth bar widths on her induced and original feathers was equal to that of the second-ranked male and greater than that of the third-ranked male (Figure 6.4). The number-two-ranked and number-three-ranked females (numbers five and six in the flock hierarchy) had growth bar width ratios less than and equal to, respectively, the lowest ranked male.

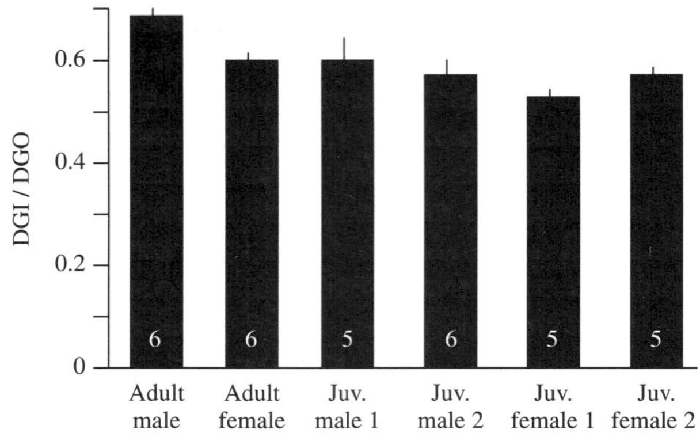

Figure 6.4 Ratio ± SE of growth bar widths of induced (DGI) and original (DGO) left outer rectrices (L6) of the two adult and four juvenile willow tits in a typical winter flock. Even though the adult female was fourth in the dominance hierarchy within a flock, the feather growth records suggest her nutritional condition was equal to or better than that of the two juvenile males, a discrepancy thought due to the protection afforded her by her dominant male mate. Sample sizes are shown in the bars. From Hogstad (1992).

Hogstad ascribed the modified foraging behavior and the relatively broad growth bars of the adult female to protection by her mate from the two juvenile males in a flock. It would be valuable to be able to follow this descriptive study with a manipulative experiment to test more powerfully his prediction about the consequence of male protection. However, it is difficult to imagine how this might be done in practice. The obvious manipulation would be to remove the dominant, adult male and then track the adult female's behavior and induced growth bar width when she was exposed to the aggression of the two juvenile males. However, we know from other work with willow and other tits that as soon as a dominant bird disappears, presumably having been killed by a predator, the second-ranked bird of the same sex abandons his or her putative mate in the flock and moves up to mate with the newly unmated dominant of the opposite sex. In a removal experiment, as soon as the dominant male was removed, the second-ranked male would likely move up to pair with the dominant female and begin shielding her from the aggression of the formerly third-ranked male.

We continue this investigation of the nutritional correlates of social dominance with reports from studies of two other titmouse species in Europe. The first project, from Spain, reinforced the conclusions of Hogstad in Norway, while the second, from England, produced quite a different result.

It can be argued that the great tit is the best-studied wild bird in the world, certainly the best-studied wild non-game bird (Figure 6.5a). From Ireland to Japan, the range of the great tit overlaps with the home-ranges of hundreds if not thousands of professional and amateur researchers, in most of their range great tits are accessible for study year-round, and they readily nest and roost in nest boxes where they can be captured easily. Finally, they are found even in small woodlands in urban areas. (Near where I am writing this, great

Figure 6.5 Great (a) and coal (b) tits, two species that have figured prominently in European studies of the nutritional consequences of dominance status. Photographs by Luis M. Carrascal.

tits are foraging in ginko trees on the Emperor of Japan's palace grounds, surrounded by the 40-million-person metropolis of Tokyo.)

Luis Carrascal and his colleagues examined the effect of social dominance on nutritional condition in Spanish great tits (Carrascal *et al.* 1998). Because they employed two study areas, one colder and with less food than the other, they also considered the effects of ambient temperature and prevailing food supply on nutritional condition. An unusual feature of their study design consisted of supplemental food being available to all birds all the time in both study areas. Noting that all previous studies of the species had taken place at more northern latitudes, they had the goal of determining the relationships among dominance status, access to food, and nutritional condition in great tits wintering in southern Europe.

One study area, El Ventorrillo, consisted of 6 hectares of coniferous forest at 1,500 m elevation in central Spain. The other, Sarriá, occupied 3 hectares of orchard and conifer grove at 100 m elevation near the Mediterranean coast. During the period of the study, the daily average temperatures at the mountain and coastal study sites were 4.6 °C and 11.7 °C, respectively. The workers sampled arthropod populations at both locations, dividing such potential tit food items into three size categories.

During the first half of November, the tits at the both sites were trapped, banded, had their left and right fifth rectrices plucked, and were then released. About 45 days later, induced feathers were taken from re-trapped birds. Feather growth measurements employed in the analysis were averages for left and right feathers, and results were reported as the ratio of induced to original measurement values. In addition to using feather growth as an index of nutritional condition, Carrascal *et al.* used ultrasound to estimate the mass of the pectoral muscle, the major flight muscle of birds, under the assumption that the more massive this muscle, the better a bird's nutritional state. They also suggested that feather growth and pectoral muscle mass may be related in that muscle likely acts as a temporary storage site for amino acids that are assembled into feather protein.

As we have seen in earlier studies, dominance relations were determined using the behavior of birds at the feeders, and the pattern detected was the same as seen in willow tits, males dominated females and adults usually dominated juveniles of the same sex.

The abundance of potential arthropod prey items was lower at the mountain site, both in total numbers and in numbers of large (>2 mm) items. Growth rate of induced feathers was not related statistically to body condition (pectoral muscle mass), but was slower at the colder, less food-rich mountain site than at the coastal site (Figure 6.6). At the mountain site, adult male great tits grew induced feathers at a significantly faster rate than did birds in the other three age/sex categories, but such a difference did not occur at the milder coastal site.

Carrascal and his co-workers concluded from their study that environmental stress, as reflected in rate of feather growth, varied according to the ability of various age and sex classes to compete for food. At the mountain site, the dominant adult males used their preeminent access to supplementary food to grow induced rectrices as fast as at the milder coastal site. Females and subordinate males in the mountains did not have such extensive access to food and grew feathers significantly more slowly than both adult males in the mountains and females and subordinate males on the coast. As was the case with the Norwegian willow tits, the dominant status of the adult males

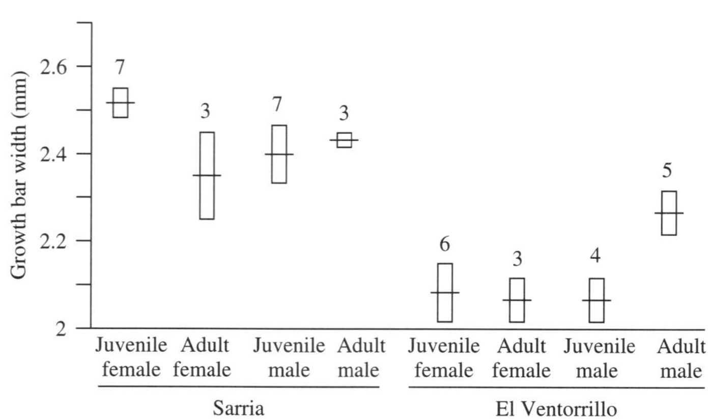

Figure 6.6 Daily induced feather growth ± SE of great tits at two locations in Spain. Such growth rates indicate that birds at the colder mountain site were in poorer nutritional condition except for adult males that presumably benefited nutritionally from their socially dominant status. Sample sizes are shown above the boxes. From Carrascal *et al.* (1998).

furnished them with the best nutritional condition among the four age/sex classes of tits at the mountain location. The less certain nutritional status of subordinate birds was shown by the females and subordinate males matching the adult males' nutritional condition on the coast, but not in the colder, less food-rich mountains.

I am dubious about one conclusion of this project and will first quote from the paper's discussion section. "Moreover, protein reserves [measured as pectoralis muscle thickness] were not consistently correlated with induced feather growth across individuals at both localities. Thus, nutritional status had no clear and straightforward effect on growth of induced feathers." This statement shows that the authors retain the bias of equating muscle mass with nutritional status. Earlier, I presented evidence that birds can increase or decrease their muscle mass voluntarily as an adaptation, and I continue to view rate of induced feather growth as the more valid of the two indices of nutritional condition. Rather than analyzing feather growth rates after controlling for pectoral muscle mass as done here, I suggest we analyze pectoralis muscle mass after controlling for feather growth rate.

While having feeders in place throughout the study period certainly aided in determining dominance status and allowed insight into birds' reliance on supplementary food, it made very problematical the valid determination of nutritional status. The distinction in feather growth and, perhaps, pectoral muscle mass between the arthropod-poorer colder mountain site and the arthropod-richer warmer coastal site was certainly diminished by the presence of supplemental food.

Finally, the scope of the project needs comment. One study site was only six hectares and the other only three hectares and yet the data from individual birds were treated as statistically independent. Surely, they were not independent and it is difficult to endorse the idea that records from these two tiny study areas are representative of mountain and coastline habitats in general.

A reversal of the positive correlation between within-species dominance status and induced feather growth comes from a study of wintering groups of coal tits in northeast England (Hay *et al.* 2004; Figure 6.5b). Hay and colleagues studied coal tits coming in the company of great tits and blue tits to a permanently provisioned feeder, the latter two species both socially dominant to coal tits. The coal tits were not living in small stable groups, as do willow tits, but were part of a larger, less rigidly structured assemblage, much of which had undoubtedly been attracted to the permanent source of supplemental food. Because of the lack of stable flock structure, the researchers used a statistical method to determine the dominance status of each bird relative to the other birds with which it interacted at the feeder.

During the winter, the central tail feathers of 18 coal tits were plucked after the birds had been captured at the feeders. Between March and June, 13 of these birds were recaptured, so the analysis was based on feathers from just 13 birds. Growth bar widths, averaged for each pair of rectrices, were determined on induced and original feathers and the usual method of

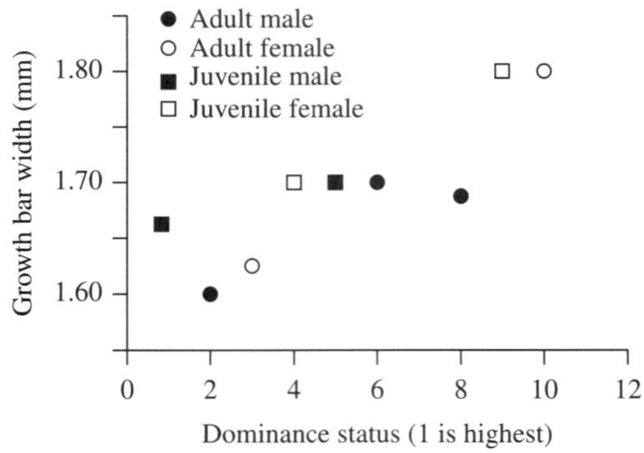

Figure 6.7 Relationship between intraspecific dominance status of coal tits and growth rate of induced feathers. From Hay *et al.* (2004).

controlling for bird size was employed using the proportion of the original growth bar width represented by induced growth bar width.

Hay and colleagues found that the relationship between coal tit within-species dominance rank and growth bar width was not only statistically significant, but was negative rather than positive. That is, the more socially dominant a coal tit, the more slowly it grew its induced feathers (Figure 6.7). (Incidentally, it is not clear why there are only nine points in Figure 1 of Hay *et al.* 2004, from which Figure 6.7 is taken, when they stated that they plucked the induced feathers from 13 birds.)

Do the results of Hay *et al.* reflect some fundamental difference between the biology of coal tits, on the one hand, and willow and great tits, on the other? Several extenuating circumstances suggest maybe not. First, the birds had access to food throughout the winter, and that situation quite likely elevated the subordinates' nutritional condition above what it would have been in the absence of supplemental food. For example, at the coastal site in the Spanish great tit study (Carrascal *et al.* 1998), the supplemental food was apparently sufficient to raise the subordinates' nutritional condition to the level of the dominant male. In Norway in the absence of supplemental food, subordinate willow tits had much poorer nutritional condition than the dominant as indexed by feather growth rates.

Coal tits are the smallest and most socially subordinate of the British tits. In the circumstances of the Hay *et al.* project, even dominant coal tits were far from the most dominant tit using the feeder. We are not told how many great tits and blue tits used the feeder, but virtually every one of them would have dominated the dominant coal tit. Such a situation is very different from that in any of the other tit studies reviewed so far. In Spain, there may have been tit species other than great tits using the feeder, but great tits dominate all other tit species there, making the dominant great tit the dominant tit at both the mountain and coastal sites. In Hogstad's study area in the mountains of central Norway, willow tits are the only species of tit present so the dominant willow tit was the dominant tit. For a true comparison, we would need to find a situation in Britain or elsewhere

where the relationship between dominance status and feather growth rates of coal tits could be examined in the absence of the confounding effects of socially dominant species. Such a situation may not exist.

In interpreting their results, Hay *et al.* brought up a matter that would be worth exploring further. Noting that the presence of the feeder may have caused coal tits, particularly dominant coal tits, to participate in an unnaturally large number of aggressive interactions with an artificially dense population of other tits, the authors suggest that the result for dominant coal tits might have been elevated metabolic rates. Dominants may also have been adversely affected by low levels of immunocompetence caused by high concentrations of testosterone. Relationships among all of these factors and nutritional condition are currently unknown and clearly attackable experimentally with current methods.

The last contribution by ptilochronology to the study of intraspecific dominance is provided by Daniel Cristol and his students (Jenkins *et al.* 2001). These workers restricted their study to wintering male white-throated sparrows, creating 14 four-sparrow social groups, each in a separate aviary. From the results of behavioral observations, each sparrow in each flock was assigned a dominance score based on the proportion of flock-mates dominated. It turned out that two behavioral measures of dominance were instructive. Subordinates spent more time than dominants waiting for another bird to leave the food dish, and made significantly more visits to the food bowl than did dominants.

Thirty-six days were allowed between the plucking of original and induced feathers, the R5 and L5 rectrices in this project. During that period, each flock of four birds was provided daily with a food ration that might be described as egalitarian in terms of quantity, but not quality. Thirty-five grams, in the aggregate, of corn, millet, sunflower hearts, thistle seeds and turkey-starter mash, and eight mealworms were provided in one dish per aviary. As a result, there was enough food of less preferred types, such as corn, to last through the day, but preferred types of food, the mealworms, sunflower hearts, and thistle, which contained a higher proportion of fat and/or protein, were gone within 1 hour of being introduced.

As it turned out, the various flocks differed in how despotic the dominant bird was, which the researchers thought might be related to the difference in feather growth measures between most dominant and most subordinate flock-mate. To account for this variation, they used the difference between the dominance scores of the most and least dominant bird to characterize each flock and compared that value to feather growth rates. At first glance, their results were equivocal. Across the 14 flocks and the 36-day feather re-growth period, there was a significant relationship between the degree of despotism of the dominant bird and how much longer his induced feathers were in total length than were those of the subordinate (Figure 6.8).

No statistically significant relationship was found between dominance rank difference and difference in growth bar width, but this lack of significance is likely due to the difference that was present being too small to be detected

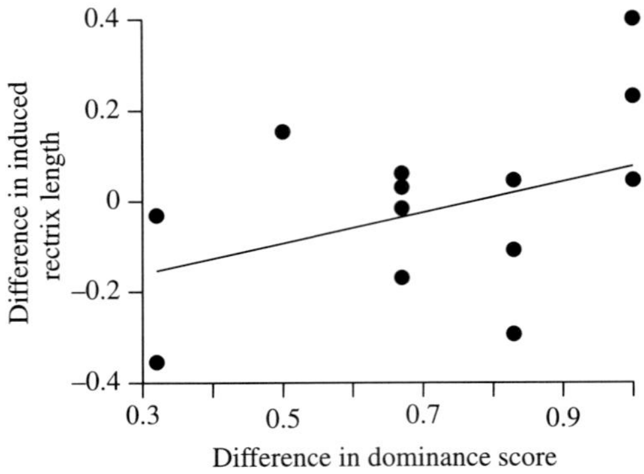

Figure 6.8 The difference in size-corrected induced feather length is positively related to the difference in dominance score between the most and least dominant white-throated sparrow in captive flocks of four birds. From Jenkins *et al.* (2001).

statistically. We know that all birds had 36 days to grow their rectrices and so total length is the accumulated growth over 36 days. By contrast, growth bar width is calculated per day. Thus, any difference in growth bar width could be about 36 times less than the difference in total length. Coupled with the fact that all birds were on a quantitatively *ad libitum* diet for the duration of test, this lack of significance for growth bar width suggests that differences in daily rectrix growth existed, but were too small to be detected.

6.2 *Interspecific social behavior*

In addition to the application of ptilochronology to within-species questions, two projects, both from my own laboratory, have extended questions about nutritional condition to multi-species assemblages. The two studies in question both stem from our group's long-standing interest in the ecology and behavior of permanent-resident woodland birds in winter. During that season of the year in Ohio, members of four to seven species move together in an organized flock through woodland while looking for food. Such flocking, termed mixed-species flocking, has been viewed in terms of advantages and disadvantages to the various species involved (Barnard and Thompson 1985). The main arguments are that the benefits of flocking have to do with sharing vigilance for predators. Being vigilant means looking around the habitat in case a hawk or other predator might be approaching. By sharing the overall task of being vigilant with flock-mates, the argument goes, any one animal will have a larger percentage of its own time available for looking for food itself. Thus, flocking allows an animal to forage for food more efficiently than if it fed alone.

The logic just outlined applies to both single-species and multi-species flocks. There is a corollary positing why foraging with other species should be more adaptive than foraging just with conspecifics. The argument is as follows: if a bird looks for food only with conspecifics, every one of its flock-mates will be a close competitor for what food is available. Such conspecific flock-mates will tend to be looking for the same types of food items in similar places. In other words, they will have the same ecological niche. Thus, in order to get the vigilance protection of a flock, an animal may pay a high price in terms of food competition. The argument continues that foraging with similar, but different, species alleviates the difficulty about food competition because individuals of different species look for different types of food in different places in the habitat; they have different niches. At the same time, different species flocking together can share vigilance duties to their mutual benefit.

Such an argument makes good sense in a Panglossian world where all birds cooperate and no one takes advantage of others. But that is not how the real world works. Just as there are dominance hierarchies within species, so there are among species. The coal tits in the study of Hay *et al.* (2004) were subordinate to both great and blue tits in mixed-species flocks. We know from work beyond the scope of this book that in mixed-species flocks, dominant species can displace subordinate species from preferred foraging locations in a woodland, that dominant species can supplant subordinates from locations where the latter have found food, and can force subordinates to drop food items they are carrying. So, the question, why forage with other species, seems rather straightforward for dominant species; they can reduce their vigilance for predators and at the same time realize a foraging benefit from displacing, supplanting, and stealing food from subordinates. The question is more difficult for subordinates. They are displaced from their foraging substrates and, while they may get help from dominants with vigilance for predators, they may need to remain vigilant for supplanting attacks by dominant flock-mates.

In the mixed-species flocks we have studied in woodlands of the American Midwest, there is a clear-cut dominance hierarchy among the four principal species. In order from most to least dominant, the species are downy woodpecker, white-breasted nuthatch, tufted titmouse, and Carolina chickadee. These four species remain more or less continuously in each other's company, foraging together throughout the winter day. They are occasionally joined by other species such as red-bellied woodpeckers, hairy woodpeckers, brown creepers, and even pileated woodpeckers once in a while. While these other species are less common, the four main species are abundant enough and easily enough captured to make good subjects for study of the nutritional consequences of mixed-species flocking.

In the first of the pair of studies on the subject, David Cimprich and I examined the nutritional consequences of mixed-species flocking for the least dominant species, the Carolina chickadee (Cimprich and Grubb 1994). Presumably, this chickadee could either benefit nutritionally from sharing vigilance with flock-mates of the three dominant species, or it could suffer a nutritional cost of being displaced and stolen from.

In the rural American Midwest, the landscape tends to be dominated by row-crop agriculture with scattered woodlots remaining from the much more extensive forest that was cleared during the nineteenth and twentieth centuries. We have capitalized on these little woodlots in experimental designs, treating them essentially as giant Petri dishes that can be randomly assigned to treatment and control groups.

Tufted titmice are a close relative of chickadees, are vigilant to about the same extent, and have very similar foraging niches. Thus, we predicted that if there is a nutritional cost to chickadees foraging with titmice, in woodlands where we removed all titmice at the beginning of the winter, chickadees should grow induced feathers with wider growth bars than should chickadees in woodlands where titmice had been left in place. If there is no cost, there should be no difference in chickadee feather growth in the two types of woodland. Finally, if chickadees benefit nutritionally from the presence of titmice, wider chickadee growth bars should occur where the titmice have not been removed.

We also predicted consequences of titmouse removal not directly measurable with feather growth bars. If titmice normally displace chickadees from preferred foraging sites, removing titmice should result in chickadees shifting toward the titmouse niche. If titmice normally share vigilance for predators with the chickadees, in the absence of titmice, chickadees should devote a larger proportion of their time to being vigilant. Since our investigation was conducted outside of the breeding season, survival was the most directly fitness-related quantity available for study. If chickadees experience an increased survival cost as a result of a competitive interaction with titmice, then in the absence of titmice, chickadee survivorship should increase. Alternatively, if flocking with titmice provides a net benefit, then chickadee survivorship should decrease when titmice are removed.

Over the course of four winters, we used five different woodlots for the project. In the first winter, we recorded observations in a single woodlot as a control (no titmice were removed), while in the following three winters, we employed both control and experimental woodlots. Throughout the experiment, four woodlots received both treatments. In each case, they served first as a control and then, in the following winter, as an experimental. In this design, control woodlots would not be influenced by any residual effects of titmouse removal that had occurred during the previous winter.

In December of each winter, we caught at temporary feeders, color-streamered, and plucked the outermost right rectrix from chickadees in control and experimental woodlots. All titmice captured in experimental woodlots were banded, transported 50 km away, and released amidst the bird feeders of suburban Columbus. None of these were observed again at any capture site. We then emptied the feeders. From late December through early February, we observed chickadee foraging and vigilance, also recording air temperature and wind speed during each woodlot visit. In late February we restocked the feeders and caught and plucked the newly grown induced feather from surviving chickadees.

We considered a chickadee to have survived the winter if it was observed during the recapture period. During the project, no chickadees were ever observed to immigrate into the study woodlots and no marked chickadees were observed in neighboring woodlots. Therefore, all chickadee disappearances were treated as deaths rather than emigrations.

Try as we might, we were not able to remove all the titmice from all the treatment woodlots; some were just too trap-shy. However, we were able to remove the great majority of them. In our regular survey routes through woodlots, we noted 3.0 titmice observations per km in control woodlots, but only 0.4 titmice per km in treatment woodlots. Looking at the reduction in another way, 39% of 272 chickadee-flock observations in control woodlots included at least one titmouse compared to only 11% of 171 chickadee-flock sightings in experimental woodlots.

In the almost-total absence of titmice, chickadees tended to forage more in the titmouse niche; some of these trends were statistically significant while others were not. In the experimental woodlots, chickadees foraged more on the ground, more high up in trees, more on branch sizes and substrate categories (e.g. trunk, branch, twig) favored by titmice, and more on dead wood.

Mean chickadee vigilance rate trended higher in the absence of titmice, but the trend was not significant.

In the multivariate analyses of feather growth, values of original feathers were treated as covariates. Both male and female chickadees grew their feathers at a faster rate in the absence of titmice (Figure 6.9). The difference between control and treatment birds was greater for males (11%) than for females (6%), perhaps because female chickadees are socially subordinate to

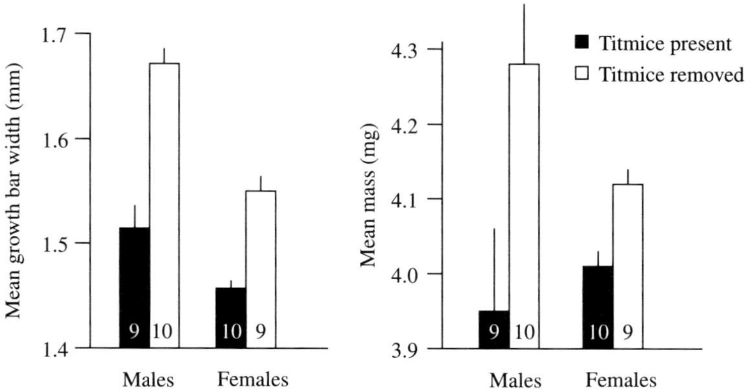

Figure 6.9 Mean ± SE growth bar width and mean ± SE mass of induced Carolina chickadee rectrices grown either in the presence of tufted titmice or in woodlands from which titmice had been removed. These means were adjusted statistically to cancel out the effects of variation due to growth bar width and mass of original rectrices from the same birds. Sample sizes are shown in the bars. From Cimprich and Grubb (1994).

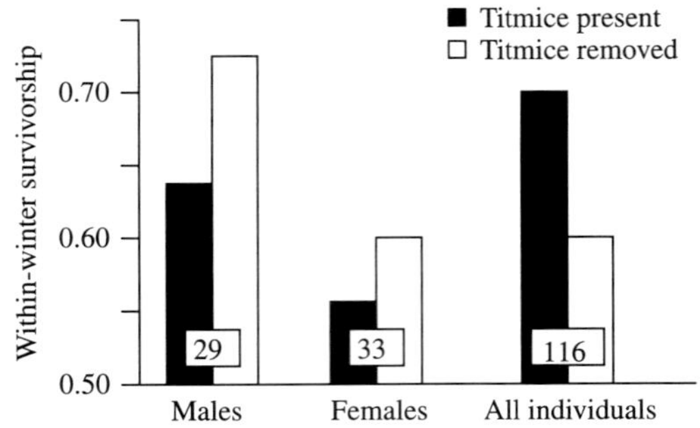

Figure 6.10 Within-winter survivorship of male, female, and all Carolina chickadees in the presence of tufted titmice or after the titmice had been removed. From Cimprich and Grubb (1994).

males of the species so the removal of titmice would still leave them dealing with dominant chickadees. Analysis of the mass of feathers produced similar results, greater mass in the absence of titmice and a larger difference between treatment and controls in males (8%) than females (2%).

In both experimental and control settings, the over-winter survival rate of males was higher than of females. There was also a trend for survivorship in both males and females to be higher in the absence of titmice, but these trends did not reach statistical significance (Figure 6.10). Fifty-four of the 116 chickadees in the project had wing lengths intermediate between those clearly of females (<61 mm) and males (>64 mm). When these birds were added to the sample and the survivorships of all experimental and control birds, regardless of sex, were calculated, just the opposite result occurred; chickadees had higher survivorship in the presence of titmice than in their absence (Figure 6.10). The project was done before the days when we could sex chickadees unambiguously using molecular genetics, as we do routinely nowadays.

This study of interspecific dominance seemed to show reasonably clearly that chickadees paid a nutritional cost by foraging with titmice because they were displaced away from their preferred and presumably most efficient sites for looking for food. If the trend toward reduced vigilance in the presence of titmice was real, the resulting increase in foraging time did not make up nutritionally for the forced shift in foraging niche.

The effort to connect nutritional condition to survivorship left us in a muddle. Why known males and known females should have higher survivorship in the absence of titmice, but all chickadees combined should trend strongly the other way is a mystery. As always in such cases, we can invoke sampling error in the sense that with an alpha level of 0.05, one in 20 cases of proving the null hypothesis wrong will, itself, be wrong.

Before leaving this set of data, we might think about what the chickadees' actual behavioral strategy might be regarding foraging with titmice. We will see in the next example that it seems to be in the interest of dominant species to stay in the company of subordinate species, for all the reasons outlined at the start of this section. But what choice do subordinate species have? Conceivably, they could try to avoid the dominants. However, it may actually be more costly nutritionally to try continually to keep away from dominants, sacrificing foraging time in the process, than to just pay the price of flocking with the dominants.

The next and last project in this chapter considers the other side of the interspecific dominance coin, the side occupied by dominant species. In our study sites in Ohio, downy woodpeckers and white-breasted nuthatches are socially dominant to Carolina chickadees and tufted titmice. There is another distinction between the chickadee and titmouse on the one hand and the woodpecker and nuthatch on the other. The chickadee and titmouse belong to a class of species termed nuclear species. Nuclear species tend to be active, noisy, and at the forefront of movement of a mixed-species flock through its habitat. By contrast the woodpecker and nuthatch belong to a category of flock participants called satellite species. These are quieter, less active and tend to lag behind the nuclear species in flock movements. In the case of the downy woodpecker and white-breasted nuthatch, the satellite species are also socially dominant to the nuclear species, but this is not always so. For example, in our Ohio woodlands, brown creepers are a satellite species that is socially subordinate to the chickadee and titmouse.

Andrew Dolby and I were interested in exploring the impact of mixed-species flocking on the behavior, nutritional condition, and survivorship of the two common satellite species in our woodlands, the downy woodpecker and white-breasted nuthatch (Dolby and Grubb 1998). We reasoned that if these dominant species derived nutritional benefits from flocking with the nuclear species, then if we removed the chickadees and titmice from a mixed-species flock, we ought to see several things about the satellite species change. First, assuming that satellite flock members in the absence of nuclear species must sacrifice foraging time to increased vigilance, as Kim Sullivan has shown (Sullivan 1984*a*), in the absence of chickadees and titmice, the woodpeckers and nuthatches should attempt to compensate for the reduction in food intake rate by adjusting their foraging behavior to reduce heat loss to the environment. Such heat loss would require increasing metabolism to keep the temperature of the body high and constant. We performed this experiment in central Ohio using woodlots isolated by surrounding agricultural fields. Owing to differences in exposure to wind and sunlight, microclimatic conditions can vary greatly within such woodlots. Furthermore, small birds apparently sometimes reduce their metabolic costs in winter by selecting foraging sites exposed to low wind and high solar radiation (Wolf and Walsberg 1996). If the net rate of energy intake (i.e. food intake rate minus metabolic cost) is reduced in satellite species when nuclear species are removed, then woodpeckers and nuthatches in the absence of the two nuclear species should compensate by foraging in areas more sheltered from wind and more exposed to solar radiation. Specifically, they should forage farther from the windward edge of a woodlot and closer to the "sunward" edge. They should also choose to look for food on larger diameter trunks and branches to reduce their exposure to the wind. (These predictions assume that both predation risk and food availability vary independently from exposure to wind and solar radiation at foraging sites used throughout the winter.)

Dolby and I were also aware that Kim Sullivan had shown by playing back recordings of chickadee and titmouse vocalizations that downy woodpeckers were parasitizing the vigilance of these core species (Sullivan 1984*b*). Specifically, when played the vocalizations that presumably made a woodpecker think that it was near the core species, that woodpecker would reduce the amount of its own vigilance. We therefore predicted that vigilance levels in the two satellite species would increase in woodlots from which chickadees and titmice had been removed. Of course, such an increase in vigilance could result from either a simple effect of having fewer birds in the flock, as we saw in the section on intraspecific dominance, or because chickadees and titmice act as sentinels, uttering alarm calls which are parasitized by satellite flock members.

If flocking with subordinate core species increases the energy intake rate of dominant satellite species, woodpeckers and nuthatches should exhibit reduced nutritional condition in woodlots from which the core species have been removed. We tested this prediction using ptilochronology.

Finally, if foraging with nuclear species confers to satellite species the fitness benefit of increased survivorship, woodpeckers and nuthatches should exhibit decreased survivorship in woodlots from which the chickadees and titmice have been removed. Such a mortality effect could occur due to increased predation, increased starvation, or both.

We conducted the experiment in Ohio woodlots dominated by oaks, hickories, sugar maple, and American beech. Each woodlot was completely surrounded by cultivated agricultural fields and contained one isolated mixed-species flock. Such flocks were never observed to cross into neighboring woodlots during the study period. In addition to chickadees, titmice, downy woodpeckers, and nuthatches, the woodlots usually contained one or two red-bellied woodpeckers and, occasionally, a hairy woodpecker or brown creeper.

Sixteen woodlots were used in the experiment. Before each of the two field seasons, four of the eight woodlots to be used that year were randomly assigned to the control and four to the removal group. Control woodlots averaged 5.3 hectares in size and removal woodlots averaged 5.5 hectares. Each woodlot was used just once. Before entering a woodlot on each visit, we recorded wind speed and air temperature in open fields nearby.

During the two weeks on either side of January 1, we caught downy woodpeckers and nuthatches at temporary feeders, color-streamered them and pulled two rectrices, L5 and R5 in woodpeckers, L6 and R6 in nuthatches. Chickadees and titmice in control woodlots were similarly marked, but those in treatment woodlots were removed and released approximately 50 km away in the "bird feeder belt" of the northern Columbus suburbs. We were only partially successful in re-catching birds in late February to collect the induced feather. (These larger species are less dependent on the feeders for energy than are the smaller chickadees and titmice, and so were harder to persuade to enter a trap after having been caught once earlier in the winter.) We averaged records for right and left feathers when calculating feather growth bar width, total length, and mass. All feather measurements were conducted in a blind fashion. Specifically, a third party assigned random identification numbers to feathers so that the measurer did not know whether a given feather came from the treatment or control group.

We calculated survivorship from sightings of individual birds during woodlot visits. If a marked downy woodpecker or nuthatch disappeared from a woodlot and was not found during surveys of neighboring woodlots, we ascribed the disappearance to mortality. On two occasions, we observed unmarked nuthatches enter study woodlots only to be immediately driven out by the color-marked resident pair. These observations strengthened our assumption that mortality was responsible for the cases when a marked nuthatch was replaced by a new individual.

We went to a very great deal of trouble to be able to treat the woodlot, rather than the bird, as the primary sampling unit, that is, the unit of measurement thought to be statistically independent. We were able to conclude that records from different woodlots were truly independent because the birds in any two different woodlots did not interact in any way. Still, there is some chance that more than one woodlot was visited by the same Cooper's hawk and, if so, one could conceivably argue that the two woodlots were not actually independent. (To carry this sort of purist statistical argument to its conclusion *ad absurdum*, one could argue that since any two woodlots were on the same planet, they could not be considered statistically independent, for example, because they were both warmed by the same sun. This is the sort of thing academics sometimes get bad reputations for arguing about!)

As with the titmouse-removal experiment mentioned earlier, we were not able to remove absolutely all of the desired removees. Except for two titmice during the first winter and one chickadee during the second, all parids (a word derived from the Family name of titmice and chickadees, Paridae) were removed from the eight experimental woodlots during the initial capture period. The average number of parids removed per woodlot was 5.6. Data were recorded for 21 woodpeckers and 13 nuthatches in 1996 and 25 woodpeckers and 16 nuthatches in 1997.

For male and female downy woodpeckers, no exposure-related variable differed significantly between control and treatment woodlots. However, the reduction in wind speed from woodlot edge to foraging site approached significance in treatment female woodpeckers, as did the reduction in foraging height. These differences were in the predicted directions. No significant differences in foraging-exposure variables were found for either nuthatch sex. Foraging heights tended to be lower for male woodpeckers and nuthatches during the second winter of the project, probably because two woodlots used during that year contained a higher proportion of young trees than the other woodlots used in the experiment.

Vigilance rates for all species/sex categories differed significantly between control and treatment woodlots. Male nuthatches exhibited the most dramatic response to the removal of parids: engaging in 5.4 and 10.1 vigilance bouts per minute, respectively, in the presence and absence of chickadees and titmice.

Because of the trap-shyness of many birds, only for male nuthatches were enough induced feathers collected to conduct an analysis of nutritional condition using ptilochronology. Induced feathers were collected from male nuthatches in all eight control and in five treatment woodlots (the low latter number because of presumed mortality; see later). The samples of induced feathers obtained from each of the three other species/sex categories were less than or equal to three for the 2 years combined. Feather mass and length did not differ significantly between male nuthatches in treatment and control woodlots, but mean growth bar width differed significantly between the two groups and, as predicted, was lower in the absence of parids (Figure 6.11).

Assuming, again, that birds have been selected to regenerate feathers as quickly as possible and with the greatest structural integrity, given body maintenance constraints, the pattern of induced feather growth in male nuthatches provides evidence that nuthatches may have suffered nutritionally when parids were removed.

Several mechanisms might have led to the reduced nutritional status of male nuthatches in treatment woodlots. First, the demonstrated increase in their vigilance levels might have been sufficient to reduce their net rates of energy intake. Second, copying the foraging behavior of parids could be of sufficient benefit to produce increased feather growth rates. Finally, while an inability to steal food from the socially subordinate parids may have contributed to reduced net energy intake rates in the treatment woodlots, no cases were observed of nuthatches stealing food in the control sites.

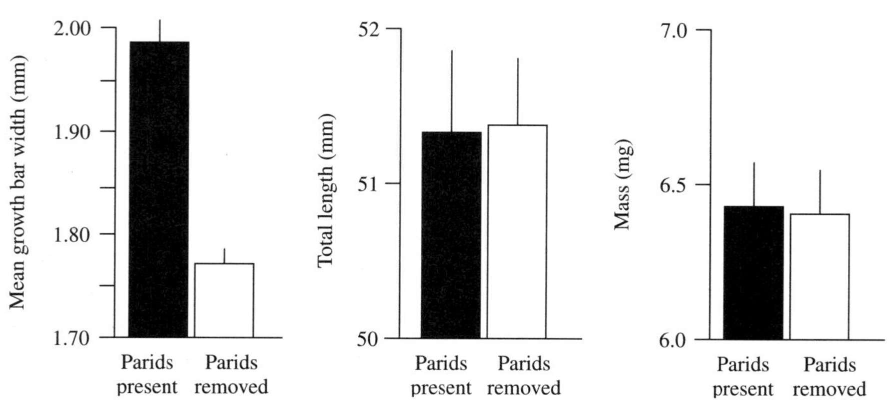

Figure 6.11 Mean ± SE growth bar width, feather length, and feather mass for male white-breasted nuthatches with parids (chickadees and titmice) and in the absence of parids. N, the number of woodlots, was eight in the control group and 5 in the treatment group. From Dolby and Grubb (1998).

No downy woodpecker mortality occurred in either the treatment of control groups. For nuthatches, no disappearances were recorded among females in either experimental group, or among control-group males. However, three males in the treatment group disappeared during the winter and were replaced by new males. Although it is not possible to be sure that these males died, they were not found during surveys conducted in neighboring woodlots.

Once in possession of a territory, a male nuthatch seldom, if ever, leaves it. The apparent mortality in treatment males (three of eight) was close to being significantly greater than that in controls (none of eight) and was in the direction predicted if foraging with parids confers a survival benefit to satellite flock members. Even if mortality was statistically greater in treatment woodlots, it would still not be possible to determine whether individual deaths were caused by predation, by increased starvation due to compromised foraging effectiveness, or by both mechanisms.

Considering together the present results and those of the titmouse-removal project reported earlier in which chickadees benefited nutritionally from the absence of titmice, the type of fitness benefit derived from foraging with other species may depend on social dominance status. Within the mixed-species flocks that we have studied, the dominant satellite species may garner both anti-predation and foraging benefits. By contrast, the most subordinate nuclear species, the chickadee, may derive anti-predation benefits, but incur a foraging cost.

6.3 *Summary*

Ptilochronology has given us insights into the nutritional consequences of behaving socially with members of the same and other species. In broad-brush terms, the studies reviewed here have found that it is advantageous for birds to forage socially with members of the same species rather than exist alone and that, within a species, social dominants enjoy better nutritional condition than subordinates. In mixed-species assemblages, the nutritional condition of dominant species improves, while subordinate species may pay a nutritional price for the increased vigilance of flocking.

Most of the studies cited have employed the preferred methodology of using data from induced rather than original rectrices and some have randomly assigned birds, or habitat patches, to control and treatment groups. A continuing problem, and one that will probably permanently plague this and other experimental field studies of birds, is the enormous amount of labor per data point. As a colleague of mine who studies spiders once put it, "Why do you study birds rather than spiders?" One answer is that birds do things that spiders do not do, for example, actively moving through the environment in social foraging groups.

7 *Individual quality*

7.1 *Sexual selection and natural selection*

With this chapter, we begin reviewing a series of ways that ptilochronology has been employed in studies of reproductive behavior. We begin by focusing on aspects of mate quality, followed by projects dealing with reproductive effort, nestling condition, and prolonged brood-care.

Ever since Darwin (1871), biologists have been fascinated by the interplay between sexual selection and natural selection. Sexual selection is selection on the characters of one sex by the mating preferences of the other sex. For example, the splendid upper tail coverts of peacocks have been selected for by sexual selection. That is, they have been selected for by the mating preferences of peahens. Because of such a preference, males with relatively longer, more showy covert feathers copulated with and fertilized more peahens and, therefore, left more genes to the next generation. Through evolutionary time, presumably, this sort of directional sexual selection produced the magnificent plumage of modern peacocks.

Natural selection, on the other hand, is usually defined as selection on an organism's characteristics by all forces in the environment other than sexual selection. (Actually, if pressed, one would have to admit that sexual selection is actually but one variety of natural selection.) In all known cases, sexual selection seems to have been checked eventually by countervailing natural selection. In the peacock, presumably, sexual selection for longer, heavier coverts has been stopped by natural selection opposing longer coverts. Perhaps, coverts any longer make a peacock too slow to escape predators or, perhaps, production of even longer, heavier coverts would leave insufficient energy and nutrients for body maintenance.

We imagine sexual selection pushing characters up against the barrier of natural selection. Animals that have such sexually selected characters are said to have handicaps. Handicaps are just what the name implies, they are characteristics not favored by natural selection. All else being equal, a peacock with longer, heavier feathers possesses a handicap making him more subject to predation or starvation. Amotz Zahavi first articulated the idea that females selecting mates should prefer males with big, showy, metabolically expensive ornaments because the fact that they possess such ornaments and are still alive and in good nutritional condition must mean they are of better quality than males that possess smaller less showy ornaments. Zahavi labeled this idea the Handicap Principle (Zahavi 1975). Females should select males that are able to bear big handicaps and still be in good condition. Why? Because, assuming the handicapping trait is heritable, females mating with handicapped males should produce sons able to bear big handicaps and therefore be attractive to females, leaving more offspring (and grandchildren of the choosy female) to the next generation.

The final aspect of the handicap principle is that in order for a handicap to be selected by sexual selection, it needs to be truthful. That is, if the presence of a big handicap can be faked by males that are not in good nutritional condition, then mate-seeking females will be fooled or confused and handicapped males will not leave relatively more offspring. In the parlance of behavioral ecologists, the signal must be an honest signal in order to have been favored by sexual selection (Kodric-Brown and Brown 1984).

We now know that females are not passive receivers of copulations; they are actively making choices about which males other than their social mate to copulate with. Such copulations outside of the social pair-bond are called extra-pair copulations (EPCs) and extra-pair fertilizations (EPFs) result when EPCs produce offspring. Because some males are the fathers of nestlings in the territories of other males, we now see that sexual selection can be even more important than we had previously thought. The fitness differences among males of the same species can be greatly increased if some males are fertilizing eggs in the nests of other males as well as in their own nest.

These two lines of thought, the handicap principle and the discovery of widespread EPCs and EPFs, have worked together to produce many projects focused on the characteristics of birds, usually males, that are honest signals of male quality, signals that can be used by females for both social mate selection and genetic mate selection via EPCs. We start a review of the contributions made by ptilochronology with a series of papers focused on aspects of plumage, itself, as potentially honest signals of bird quality. By quality, I mean the ability to confer genes to the next generation by attracting one or more mates per breeding season, by surviving for many breeding seasons, or both. The first paper in the series focuses on the possibility that growth bar width itself is an honest signal of quality while the next four reports concern work correlating some arbitrary (but honest) plumage signal with feather growth evidence of high nutritional quality.

7.2 *Growth bar width as an honest signal*

Yoshihide Takaki and his co-workers studied the reproductive biology of a Japanese songbird, Styan's grasshopper warbler, on a small islet in the Sea of Japan near the main Japanese island of Kyushu (Takaki *et al.* 2001; Figure 7.1). On that 2.5-hectare tabletop scrap of land, 25 to 30 social pairs of this socially monogamous, territorial, migrant species breed each year. As do many songbirds, these warblers undergo a full molt of body and flight feathers after breeding and before migrating in the autumn. Takaki *et al.* correlated the width of growth bars on the original R3 rectrix molted in late summer of one year with various aspects of breeding biology the next spring. In this color-banded population, because an unknown proportion of first-time breeders on the island could have grown such original R4s elsewhere the

Figure 7.1 (a) Styan's grasshopper warbler, and (b) the species' tail spread to show growth bars very apparent on all rectrices. Photograph by Hisashi Nagata.

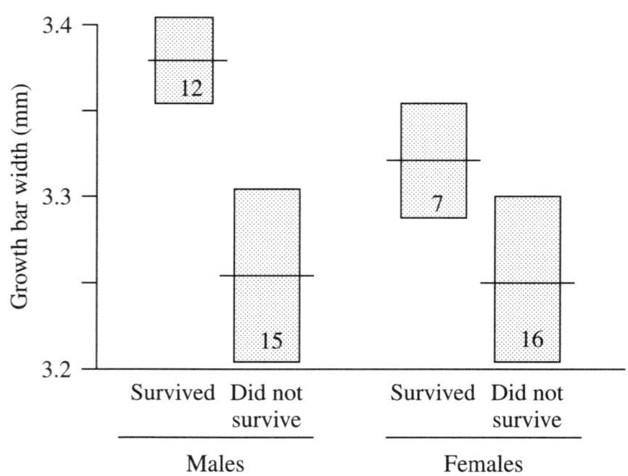

Figure 7.2 Growth bar width on original R3 rectrix and annual survivorship in Styan's grasshopper sparrow. Sample sizes are shown in the bars. From Takaki *et al.* (2001).

previous year, an uncontrolled variable, only birds known to have bred and molted on the island were included in the analysis.

For both males and females, there was a significant positive correlation of growth bar widths of individual birds between the two years of the study. Birds in better condition during one molting season were also in better condition during the second one, a hint at a genetic influence on quality. In an important positive correlation supporting growth bar width as an index of survivorship, males that survived from one year to the next (or at least returned as breeders to the same island) had relatively wide growth bars, but females did not (Figure 7.2). It is well known that among migrant territorial songbirds, males that survive are more likely to return to a previous year's breeding site than are females, and this result suggests that the warblers conformed to such a generality.

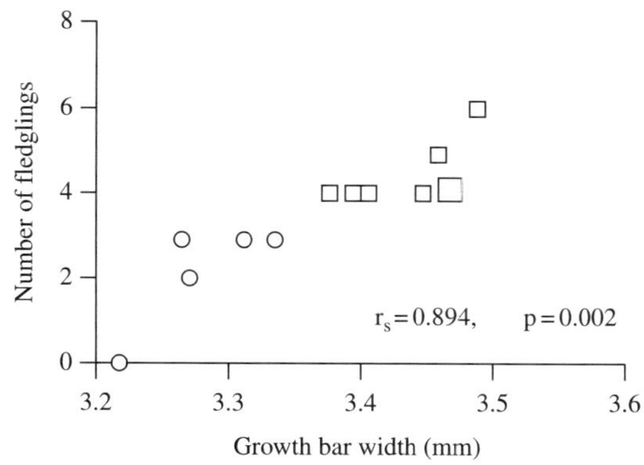

Figure 7.3 Relationship between the width of growth bars on male Styan grasshopper sparrow rectrices grown during the fall molt and the reproductive success (total number of fledglings) of those males during the next breeding season. Males producing one and two clutches during the one breeding season are indicated by circles and squares, respectively. The large square indicates two data points. From Takaki *et al.* (2001).

In males in both years and females during the second year, birds with wider growth bars were more likely to return relatively early to the breeding island in the spring. In both years, males with wider growth bars acquired a social mate earlier than males with narrower growth bars. The same was true for females, but only in the second spring of the study.

In the second important correlation from this study supporting feather growth rate as a measure of fitness, growth bar width in males was positively correlated with total reproductive success in a season (Figure 7.3). This correlation was largely due to the fact that only males with wide growth bars raised two broods, and they raised two broods because they returned early and were able to attract a female and begin breeding early. In females, the same positive tend between growth bar width and seasonal reproductive success was found in the second year of the study, but did not quite achieve statistical significance ($P = 0.06$). An insufficient sample size prevented analysis for females during the first breeding season.

One might suggest that it is territory quality rather than male quality during year 1 that determines the width of growth bars on a breeding male's feathers in year 2. The authors of the study argue against this idea as follows: food abundance correlates highly with territory size, but there was no correlation between a male's territory size in year 1 and the width of his growth bars, so food abundance, *per se*, may not influence growth bar width as much as the condition of the male coming out of the year 1 breeding effort.

Although this is only a descriptive study, it seems clear that human beings could use the width of feather growth bars to make inferences about the nutritional condition of these birds during the post-breeding molting period. Furthermore, a human being could predict that the possessor of wide growth bars would have a relatively high probability of surviving and reproducing successfully, the two ingredients of fitness *sine qua non*. The big question, the pivotal question, is whether female warblers use the same criteria when selecting males with which to mate socially. Are females even aware that growth bars exist on males, let alone scrutinize their widths? Furthermore, if a female is socially mated to a male with narrow growth bars, does she seek EPCs with neighboring males with wider growth bars?

We know that males that attract females early on are both early-arriving and have wider growth bars. Takaki and his colleagues point out the importance of manipulative experiments to tease apart growth bar width from other male characteristics as a cue to quality used by females. Imping is a falconer's term that means gluing the shaft of one feather inside the cut-off or broken-off shaft of another. The authors make the very nice suggestion of imping into the cut-off rectrices of a wide-growth-bar early-arriving male warbler the rectrices of a narrow-growth-bar late-arriving male. If females are paying attention to growth bar width on rectrices, such an imped bird should no longer attract females as social mates early in the breeding season. The reversed imping, wide-barred feathers onto narrow-barred males, should produce the opposite effect on the timing of mate attraction.

Suppose a female choosing a social and/or genetic mate does not pay attention to growth bar width on feathers to assess male quality, but does pay attention to something correlated positively with growth bar width. Assuming that growth bar width is a measure of nutritional condition and that nutritional condition is a measure of male quality, that something correlated positively with growth bar width would be an honest signal of quality. The next several papers have in common the notion that the strategy of a female is to pay attention to some aspect of a male color signal that is correlated with growth bar width. If there is a positive correlation, the conclusion is that the color signal could be an honest indicator of nutritional condition and, therefore, fitness.

7.3 *Feather pigment color as an honest signal*

Many animal species incorporate carotenoid pigments into their integument. In birds, carotenoids give feathers their reddish hue. No vertebrate can synthesize carotenoid pigments; they must be secured from the food. Experiments with both fish and birds have indicated that the extent of a male's carotenoid-based coloration is a cue used by females in mate choice. This result led to the assumption that extent of carotenoid-based coloration is an honest signal of nutritional condition. Geoffrey Hill (1990) showed in mate-choice experiments that female house finches preferred males with bright red carotenoid-based plumage. In the follow-up study reported here, Hill and Robert Montgomerie (1994) used ptilochronology to demonstrate for the first time that carotenoids can be such an honest signal usable by females in mate choice.

Hill and Montgomerie devised a composite score based on the hue, intensity, and tone of seven regions of the body to characterize the red coloration of plumage on male house finches. Such scores were obtained for each male

Figure 7.4 Three male house finches from the same population illustrate variation in coloration of the plumage. See also Plate 1. Photograph by Geoffrey E. Hill.

in three different wild populations, two in California and one in Mexico, and for males caught in Michigan and raised on supplemented diets in captivity (Figure 7.4, Plate 1). The caged males had been captured at least four months before they began the fall molt and were fed a rich *ad libitum* diet of sunflower seeds, commercial finch food, vitamins, and water.

The researchers applied a molt index to the wild populations that told them how far along in the molting process each bird was on the day it was sampled for color. At the same time that birds were scored for plumage coloration, the outermost right rectrix (R6) was plucked and stored. The analysis was based on growth bars in these original rectrices, feathers grown during the late-summer molting period at the same time the new body plumage, of varying degrees of redness, was being grown. Growth bar widths were measured as usual, and in a blind fashion so that the plumage brightness score paired with each feather was not known by the person doing the measuring.

In each of the four groups of finches, the three wild populations and the captive flock, there was a significant positive correlation between the brightness of a male's coloration and the width of the growth bars on his tail feather controlled statistically for total length of the feather (Figure 7.5). These trends held independently for both young-of-the-year and adult males.

It turned out that there was a substantial difference in average plumage color between the two California populations. Nevertheless, the average growth bar width for the two populations was the same. Hill and Montgomerie concluded from these facts that either carotenoid-bearing plants were more common for one finch population than the other, or that one finch population had a higher infection rate of parasites or pathogens that prevented the birds from depositing

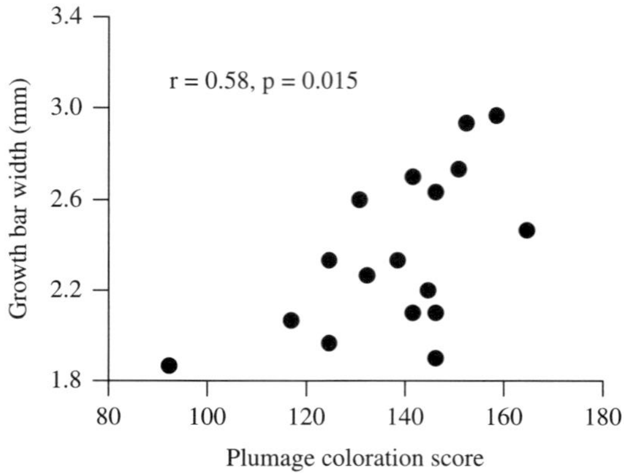

Figure 7.5 Body plumage coloration of male house finches in relation to growth bar width of an original rectrix. While results are illustrated for a population in San Jose, California, records from another population in California as well as from those in Mexico and Michigan were similar. Although the four populations differed in average coloration, within each population, relatively brighter birds had wider growth bars. From Hill and Montgomerie (1994).

more carotenoid pigments in their feathers. In any case, the conclusion remains that within either population, the pigment-nutritional-condition-quality connection would still hold because the females would be confining their plumage choices to males available within just one population.

Within each population, relatively brighter males were further along in the molt process when sampled. Hill and Montgomerie assumed that the molting process should start as soon after breeding as possible and so concluded that molt timing, as brightness of plumage, is a function of nutritional condition.

How, exactly, might it benefit a female to select red plumage and correlated wide growth bars indicating good nutritional condition of a male? How might it raise her fitness? One idea is the so-called good-genes hypothesis, which holds that if production of bright plumage is heritable and if females prefer brightly plumaged males, then a female that mates with such a male will produce sons with bright plumage that will also be preferred as mates and/or participants in EPCs. This idea rests on the notion, supported by Hill and Montgomerie's study, that bright plumage indicates superior nutritional condition. In earlier work, Hill (1991) demonstrated a second benefit to females. He found that males with brighter plumage brought more food to their mates, clearly a benefit to the female. Use of ptilochronology allowed Hill and Mongomerie to support the premises of earlier work that females selected bright males because such males were superior in nutritional condition and that such superior nutritional condition translated into at least one direct benefit for the females, more food brought to the nest during the breeding period.

The captive birds on an *ad libitum* high-quality diet grew their rectrices at a faster rate than did any of the wild birds. A small point, but it would have been nice to know that birds taken from the study populations in California and Mexico would also have grown their rectrices faster in captivity on *ad libitum* high-quality food. That would have eliminated even the smallest chance that some difference between California and Michigan birds other than that caused by diet could have been responsible for the difference in feather growth rates. To really nail down the relationship, one could use caged males randomly assigned to diets of differing quality, thereby producing birds with different brightness of red plumage along with different widths of growth bars on tail feathers. If females chose among these males produced by manipulation the same way they did with wild males, the confirming evidence would go far toward eliminating any doubt.

Hill and Montgomerie pointed out that their results nullified the primary criticism of ptilochronology by Murphy and King (1991) and Murphy (1992), that only "lethal levels of nutritional privation consistently slowed feather growth," and that "moderate or even severe sublethal privation did not produce consistent effects on feather growth," (Murphy and King 1991). The results of Hill and Montgomerie show that even among apparently healthy, free-living house finches, nutritional privation far short of lethal levels slowed feather growth.

7.4 *Feather structural color as an honest signal*

Brightly colored plumage in birds is produced by two primary mechanisms (Keyser and Hill 1999). As shown by the house finch project, one mechanism consists of incorporating pigments into a developing feather. Such pigments can be either carotenoids that produce yellow, orange, or red feathers, or melanins, which result in brown to black feathers. A second mechanism does not depend on the incorporation of pigments, instead it depends on the internal structure of feather keratin absorbing some wavelengths of light and reflecting others. To restate this idea in terms appropriate to the physics of light, "structural colors are produced by the constructive interference of coherently scattered light waves within the spongy medullar keratin matrix of feather barbs" (Prum *et al.* 1999; Doucet 2002). This mechanism gives rise to feathers blue, purple, green, and iridescent in color.

It has recently been shown that internal feather structure can also produce ultraviolet colors, colors too short in wavelength to be perceived by humans. Humans have three types of cones in the retina of the eye that, in combination, can detect electromagnetic radiation in the range of 400–700 nanometers. (1 nanometer is a millionth of a meter and the term is abbreviated as nm.) By contrast, some bird species have been shown to possess four cone cell types and the additional type is sensitive to ultraviolet wavelengths (300–400 nm; Chen *et al.* 1984; Cuthill *et al.* 2000). The presumption is that this four-cell-type system will be found in many if not most or all bird species.

We now know that in some bird species, the sexes differ in possession and distribution of ultraviolet plumage and that wavelengths of maximum feather reflectance match well the wavelengths of maximum receptivity in the avian eye. Taken together these two pieces of information suggest that structural coloration could serve as a signal produced by sexual selection. With a student, Amber Keyser, Goeffrey Hill continued his study of plumage color, including ultraviolet plumage, as a sexual signal (Keyser and Hill 1999, 2000). Their approach was similar to that with the house finches, to use ptilochronology in assessing whether plumage color could be used by females or other males as an honest signal of male quality. Although these workers had no direct evidence that blue grosbeaks, their study species, could perceive wavelengths in the ultraviolet, they assumed they could because the ability had been shown in a close relative, the northern cardinal (Chen *et al.* 1984; Chen and Goldsmith 1986).

Male blue grosbeaks vary in the color of their body plumage from bright blue to brown (Figure 7.6, Plate 2). It turns out that on the breeding ground brown males are young birds in their first breeding season. However, there is a good deal of variation among older males in the extent of and brightness of their blue body plumage. It was on the correlation between these aspects of color and nutritional condition that Keyser and Hill focused.

Males were trapped at the breeding site in Alabama, banded, had their plumage characterized in several ways, had their two outer tail feathers pulled, and were then released. As in the house finch work, the feathers taken were original feathers grown during the previous summer. Keyser and Hill

Figure 7.6 Adult male blue grosbeak. In males of this species, "blueness" of contour plumage was positively related to feather growth. See also Plate 2. Photograph by Geoffrey E. Hill.

point out that the tail feathers, uniformly brown on all males, are unlikely to serve any function in sexual selection in blue grosbeaks and so their growth rate can be considered a measure of nutritional condition that is independent from blue body plumage.

The color of the body plumage was characterized in several ways, the more important of which were the wavelength of peak reflectance and the percentage of all reflectance possible from a white surface that occurred at the reflectance peak. These and several other measures were combined to give a single score of "blueness" that could be compared with feather growth rates.

Even after the brown first-year males had been excluded, there remained considerable variation in "blueness." Figure 7.7 demonstrates this variation by comparing the percentage of maximum possible reflectance shown by the plumage of six individual males as detected by a spectrometer. Interestingly, because the human eye does not detect ultraviolet wavelengths, all birds except the one depicted by the bottom curve appeared equally blue to humans.

The several methods used by Keyser and Hill to compare plumage color and feather growth rate produced similar results, so one will suffice here. The blueness score of each bird was compared with the length of feather grown per week during the fall molt, that is, the sum of the widths of seven growth bars. Testing separately, the investigators found significant positive relationships for the blueness of both the breast and the rump with the rate of feather growth. As the males showed a higher proportion of blue rather than brown feathers on the rump, the relationship between color and nutritional condition was stronger for that region (Figure 7.8).

Because there were significant correlations between the blueness of two body regions with the measure of nutritional condition provided by feather growth rate, Keyser and Hill concluded that the structural properties of feathers producing reflected blue coloration must be nutritionally costly to produce so they could serve as honest signals of quality. However, in a statistical

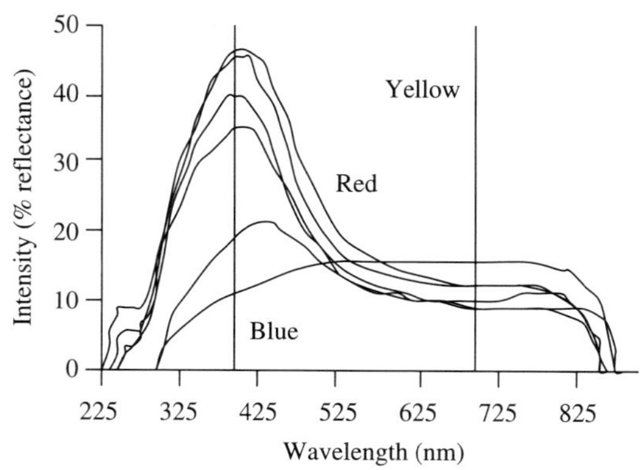

Figure 7.7 Reflectance spectra from the rump region of six adult male blue grosbeaks. Vertical lines delineate the range of human perception. Except for the minimum curve, all blue feathers appeared equally blue to the human eye. From Keyser and Hill (1999).

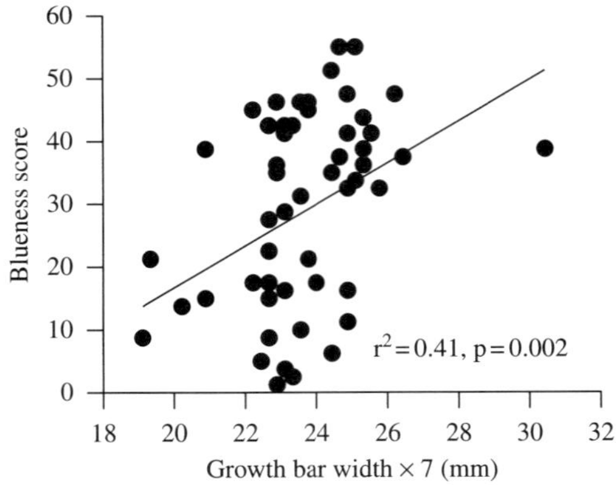

Figure 7.8 Relationship between blueness of the rump feathers of male blue grosbeaks and one week's growth (seven growth bars) of rectrices during the normal molting period. Bluer birds were in better nutritional condition during the molting period. From Keyser and Hill (1999).

sense, the blueness of the breast and rump, respectively, explained only 28% and 41% of the variance in feather growth rate. Because the correlations were not terribly tight the researchers suggest that a female may use feather brightness as only one of several cues to male quality; another cue might be vigor of singing.

Feather brightness could also function as an honest signal among males during the setting-up and maintenance of territories. Although, Keyser and Hill did not go into detail about this aspect, the likely scenario is that the brightness of a male's plumage would furnish an honest signal useful to other males in deciding whether to engage in a territorial contest.

Keyser and Hill pointed out several lines of research that could be triggered by their results. We need laboratory confirmation that varying the quality and quantity of the diet can cause the predicted variation in both plumage color and feather growth rates. The methods used in the house finch study (Hill and Montgomerie 1994) could be improved by including treatment groups with more than one level of food provisioning, perhaps "high" and "low" diets such as Jenkins *et al.* (2001) used in the white-throated sparrow study mentioned earlier. It remains to be determined with aviary choice trials whether female blue grosbeaks actually do make mate-choice decisions based on the blue coloration of males. It would also be very useful to understand exactly how a nutritionally inadequate diet affects feather structure to make a body feather less reflective in the blue and ultraviolet range. Also, can increased reflectance be tied to some male trait beneficial to females such as the higher feeding rate of more colorful house finches?

In a companion report (2000), Keyser and Hill started from the knowledge that increased blue coloration in males was correlated with better nutritional condition as shown by feather growth rates. The goal of this second study was to see if better nutrition leading to bluer plumage during the previous autumn's molt predicted differences in male reproductive performance during the following breeding season. Most of the measures of quality were those that would influence the reproductive component of fitness of a female choosing a mate.

It turned out that blueness of plumage patches was positively correlated with size of a male's territory, the density of food in his territory, and the rate at which a male brought food to nestlings in his first brood. Females choosing very blue males as social mates would receive all these as fitness benefits, indicating very high-quality males from a female's perspective.

Feather growth rates on original feathers were not only correlated with blueness of a males' plumage, they were also positively related to body size, as indicated by wing length, and to territory size. Thus, it may be that other males in a population might use blueness and size as indicators of the outcome of male-male competition with any particular male. Perhaps, for this reason, big, blue males had big territories, territory sizes they could have only acquired at the expense of other males.

A note of caution is in order on the matter of statistical independence, or lack of it. Because the field study of birds and mammals is so time-consuming per data point, we sometimes slide past recognition that much of our work violates assumptions about the statistical independence of such data points. So it is, for example, that the territory size of any given male in this blue grosbeak study is a function of his aggressive interactions with his territorial neighbors, and thus the value of his territory size depends on the values for theirs, yet all the individual territory sizes are treated as statistically independent.

Following the discovery of ultraviolet reflectance from the plumage of some birds and ultraviolet receptors in retinas of some birds, the hunt was on in many labs for further examples of these phenomena. Our research group checked, as it turned out unsuccessfully, for ultraviolet reflectance from the white plumage of chickadees and nuthatches.

In 2002, Stéphanie Doucet reported that she had found ultraviolet reflectance in the plumage of blue-black grassquits, and went on to broaden our confidence in the hypothesis that structural coloration is an honest signal of male quality. This grassquit is a small permanent-resident Neotropical songbird of open habitats. In Doucet's Yucatán Peninsula study population, the non-breeding (basic) body plumage of males is brown with blue-black patches on the upperparts and head, and the wings are blue-black with an edging of brown. By contrast, females are mostly brown, but with darker streaking on the buffy breast. Doucet argued that even though males were in basic ("non-breeding") plumage throughout her study, there was ample reason to think reflectance from blue-black feathers could have indicated male quality to females. The most important of these reasons seems to be that male grassquits perform territorial and mate-attraction displays while still in basic plumage early in the breeding season.

In February, Doucet caught males, individually marked them, took standard body measurements, and removed and stored the left outermost rectrix. For each bird, she estimated the proportion of blue-black coverage of the bird's tail, wings (including coverts), rump, breast, mantle (the back of the neck), and crown. Each bird was characterized by the mean of values from these six areas. For determining spectral characteristics of feathers, she removed three to five feathers each from wing and rump patches.

Spectral reflectance curves were constructed by determining the percentage of the reflectance of a white surface that was reflected from the wing and rump feathers at wavelength intervals of 0.3 nanometers across the spectrum from 320 to 700 nanometers. The 320–700-nanometer range is thought to represent the wavelengths of light detectable by the eyes of birds. Using Principal Components Analysis, Doucet collapsed statistically several components of feather reflectance (i.e. intensity, contrast, brightness) into one number per bird. Thus, birds with a high Principle Component 1 (PC1) score tended to be brighter, more intensely colored, and tended to reflect the most in the deep blue to ultraviolet part of the spectrum. Using the insect-pin and card technique, Doucet measured growth bars in a blind fashion, that is, without knowing from which birds they came.

The rump and wing feathers of male grassquits were most reflective in the blue to ultraviolet region of the spectrum. There was considerable variation among males in these reflectance values; feathers of the brightest birds reflected almost twice as much light as those of the dullest individuals. For the great majority of birds, reflectance was greatest in the ultraviolet range, a region of the electromagnetic spectrum invisible to humans.

Growth bar width, indicating nutritional condition of the bird when the reflective feathers of the wing and rump-patch feathers were grown, was significantly positively correlated with PC1, the aggregate measure of plumage reflectance, for both patches (Figure 7.9). This correlation led to the conclusion that at the time they grew their blue-black feathers, at least, males with higher brightness scores were in better nutritional condition. The correlation between growth bar width and mean percentage of the six areas of the body

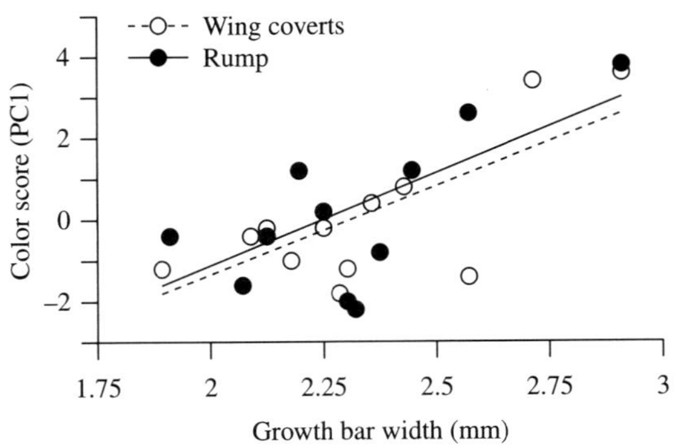

Figure 7.9 Positive relationship between plumage color and nutritional condition as indexed by growth bar width in grassquits. Correlation coefficients and p-values were $r = 0.72$, $p < 0.01$ and $r = 0.65$, $p = 0.03$, respectively, for wing coverts and rump feathers. From Doucet (2002).

Plate 1 Three male house finches from the same population illustrate variation in coloration of the plumage. Photograph by Geoffrey E. Hill.

Plate 2 Adult male blue grosbeak. In males of this species, "blueness" of contour plumage was positively related to feather growth. Photograph by Geoffrey E. Hill.

Plate 3 (a) Adult male satin bower bird, (b) Satin bowerbird bower, (c) Measuring a female bowerbird with a spectrometer. Photographs by Daniel J. Mennill.

covered by blue-black reflective feathers was also statistically significant. Structural body size, as measured by wing length, was not related statistically to overall reflectance of feathers, percentage of blue-black feathers, or growth bar width. Small males were as likely as large males to have grown many highly reflective feathers while in good nutritional condition. However, there was a nearly significant relationship between body size and brightness of the rump patch. Doucet reported that during male-male aggressive interactions, the rump patch is displayed conspicuously. Because such rump patches tended to be brighter on larger birds, birds with good nutritional condition, Doucet suggested that rump-patch color may be a badge of social status. During territorial disputes, such badges may be indicators to other males of fighting ability, reducing the need for any animal to resort to aggressive interactions that are costly in energy and can result in injury.

This work of Doucet supports the assertions of Keyser and Hill (1999, 2000) that structural plumage color is an indicator of nutritional condition and that variation in such color can be used as a cue to male quality during mate selection by females. As with the blue grosbeak, we still need confirmation from female aviary mate-choice studies with, hopefully, treatment groups consisting of males bearing plumage of manipulated reflectance. While it is not necessary for the development of evolution-level concepts, information on exactly how inadequate nutrition causes reduced structural reflectance of feathers would be quite satisfying.

Keyser and Hill (1999) have not been alone in suggesting that animals choosing mates could use an amalgam of cues indicating quality, not just some particular aspect of plumage color. The three papers in the next series all examine nutrition-dependent plumage color as well as other possible cues to mate quality. For our first example, we turn to Spanish work with that exceedingly well-studied species, the great tit.

7.5 *Carotenoids versus melanins*

There are two main categories of feather pigmentation, carotenoids and melanins. Earlier, we saw that the extent of carotenoid deposition in the body plumage of male house finches appeared to be a function of their nutritional condition. Great tits exhibit both carotenoid and melanin deposition in their feathers, and Senar and his co-workers (2003) wondered if a combination of both pigments could indicate mate quality to females. For the following theoretical reasons, they thought that such quality was more likely to be indicated by deposition of carotenoids than of melanins. Carotenoids cannot be synthesized; they must be ingested in the food. Furthermore, carotenoid pigments are known to be precursors in the synthesis of some vitamins (Olson and Owens 1998) and to play a role in the neutralization of certain harmful products of metabolism (Von Schantz *et al.* 1999). Because of these fitness-related characteristics, the extent of carotenoid deposition in feathers could well advertise an animal's nutritional quality.

Extent of melanin deposition, the authors suggest, is less likely to indicate nutritional condition because melanins can be synthesized from several common components of the diet (Fox 1976) and because no known connection exists between melanins and any aspect of general body condition (Olson and Owens 1998). Nevertheless, it had been reported that the width of a male great tit's black breast stripe was related to the quality of his parental care (Norris 1990*a*, *b*; 1993). Senar and colleagues used ptilochronology to determine if melanin deposition in the black breast stripe of male great tits was related to nutritional condition, and to compare such results with the relationship between intensity of the yellow color (due to carotenoid deposition) of other areas of the breast plumage and nutritional condition. They assumed the carotenoid deposition would be nutrition-dependent because of Hill's and Montgomerie's results with house finches.

Over the course of several years, the researchers captured 144 great tits at two study areas in Spain. The area of the black breast band was determined from digital photographs and the color of the yellow portion of the breast was characterized with a colorimeter. A nice feature of the project was recognition that the yellow and black contour feathers of the breast are found in the same feather tract so they were molted within the same time period.

At the time of capture, the second innermost pair of rectrices (L2 and R2) were plucked and had growth bars calculated in the usual fashion. Great tits molt contour plumage only during the autumn (Gosler 1993), so the black and yellow contour feathers and the tail feathers were all grown during the same time of year.

In a Multiple Analysis of Covariance, the hue of the yellow breast stripe was significantly positively related to both the width of growth bars on the tail feathers and the age of the bird, and negatively related to the number of months since the molt period (presumably because of fading). The extent of the yellow hue differed between the two study locations the two years of the project. The intensity of the yellow color was significantly related to bird age, locality, and year, while the lightness of the yellow was significantly related to sex of the bird, and year of capture. Neither intensity nor lightness of the yellow color was related to rectrix growth rate. The area of the black belly-band was not related to feather growth rate, but it was significantly related to age and sex of the bird, and to year of the study.

Senar and colleagues concluded from their analysis that yellow coloration of the breast could act as an honest signal of male quality, but that the size of the black belly-band could not as it had no relationship with nutritional condition. They mention that the main source of carotenoid pigments for great tits is the caterpillars they eat (Slagsvold and Lifjeld 1985), so a male tit with bright yellow breast plumage might signal to a female that he has a good ability to find caterpillars, not just for himself during the molting period, but to feed both her and nestlings during the breeding season. Thus, by choosing him as a mate, she would tend to increase her relative fitness.

7.6 *Multiple signals of quality*

In a pair of reports, Jodie Jawor, Randall Brietwisch, and colleagues investigated the information about an individual's quality that various male (Jawor and Breitwisch 2004) and female (Jawor et al. 2004) northern cardinal body "ornaments" might convey to potential mates. They indexed four male signals of potential use in mate choice by females. In general, the color of the upper breast plumage scored highest for color-saturated red and so was chosen as the area of the body to measure. The area and color, respectively, of the black facemask were characterized from digital images and spectrophotometer readings. Crest length was defined as the length of the longest crest feather taken from its junction with the base of the bill. Bill color was determined from comparison with color standards.

Nutritional condition during the fall molt of the flight feathers was determined by measuring in the usual fashion the width of growth bars on original feathers. Feather growth rate was not related to structural size of a bird.

For males, relationships among the various ornaments were complex. Scores for some ornaments sorted positively and some negatively with each other. However, it was clear that none of these measures was significantly associated with growth rates of the original feathers grown the previous autumn. Also, judging by the fact that the relationships were not mentioned, none of three measures of male quality, namely, feeding rate of nestlings, defense of the nest against a simulated predator, and number of nests fledging young, correlated significantly with growth bar width

Female ornaments were characterized in similar fashion although color of the underwing coverts was substituted for the color of the breast, presumably because breast feathers in females tend to be brown rather than reddish. Again, judging from the lack of mention, there were no significant correlations between growth bar width on original rectrices and any ornament measures. Three predictors of components of female quality emerged. Date of first nest and number of nests fledged were both associated significantly with color of underwing coverts. Nestling-feeding rate was positively related to facemask color score.

With regard to the predictive powers of feather growth rates as measures of mate quality, these cardinal projects seem rather uninformative. We should remember, however, that in northern cardinals the normal molt of flight feathers occurs only once a year, in the autumn, while the molt of body feathers occurs twice a year, in autumn and in late winter. Thus, on a bird-by-bird basis, aspects of breeding season body plumage (i.e. coloration of breast and facial feathers, length of crest) could reflect a late-winter nutritional condition that was either better or worse than the previous autumn. Early on in this account, our work with northern cardinals showed that the growth rate of induced tail feathers varied over the course of the annual cycle (Grubb *et al.* 1991). Therefore, comparing growth rates on original feathers as nutritional standards with records taken at other times of the year requires caution.

Yet, for whatever reason, while growth bar width of original feathers seems to be a potentially useful predictor of fitness functions in some species (Styan's grasshopper warbler, house finch, blue grosbeak, blue-black grassquit), it does not in northern cardinals.

In the next paper, we examine to what extent growth bar width is related both to aspects of plumage and to other indicators of male satin bowerbird quality potentially useful to females. This bowerbird is a denizen of Australian rainforest and, as the name implies, males build on the ground elaborate courtship arenas of sticks that they then decorate with blue feathers, flowers and berries, and with skulls of small mammals. Male bowerbirds also have patches of blue feathers on their rump, mantle, breast, and wing coverts. These blue feathers are principally reflective in the ultraviolet (300–400 nm) region of the spectrum. Stephanie Doucet and Robert Montgomerie (2003) studied relationships of bower construction and decoration and feather coloration with various indicators of male health and condition (Figure 7.10, Plate 3). We will concentrate on relationships with growth rate of original feathers.

Results were equivocal. Doucet and Montgomerie first performed a Principle Components Analysis on the mean color characteristics averaged over the four patches of color on each male. They discovered that the first principle component explained only 46% of the variation in color among

Figure 7.10 (a) Adult male satin bower bird, (b) Satin bowerbird bower, (c) Measuring a female bowerbird with a spectrometer. See also Plate 3. Photographs by Daniel J. Mennill.

males. This PC1 score was significantly related to feather growth, but inversely so. That is, according to this analysis, the brighter the male in overall coloration (UV and non-UV) the narrower the width of growth bars on the original feather. The authors go on to suggest that this puzzling result could be due to high-quality males working so hard building, maintaining, and defending their bowers during a given breeding season that they were in poor nutrition by the time the molting period arrived.

Doucet and Montgomerie found that most of the reflectance from the four color patches was in the UV (Figure 7.11), but they did not include UV brightness in the overall PC1 analysis. They analyzed UV brightness separately because it was the most variable color among males and, therefore, had perhaps the greatest potential as an honest signal of quality. They performed separate regressions of feather growth bar width on both rump and wing UV brightness. Both of these regressions resulted in significantly positive associations between UV brightness and feather growth rate. That is, males that during the previous molt had grown a rectrix at a faster rate were the same males that were brighter in the most important region of the feather-reflectance spectrum.

As always when dealing with small samples sizes, in this case 11, it is necessary to remember that results could have been prey to sampling error. Here, we have a conflict between, on the one hand, a negative relationship using as an independent variable a composite measure (PC1) that explains only 46% of the variation in the actual measured variables from which it is computed, and, on the other hand, positive relationships using as independent variables the actual measured variables in the area of the spectrum where most reflectance occurred. In their discussion, the authors did not touch on the possible difference in validity between these two types of independent variable. Even though one could argue that the UV results were more likely to be valid, the fact remains that some of the findings of this bowerbird study

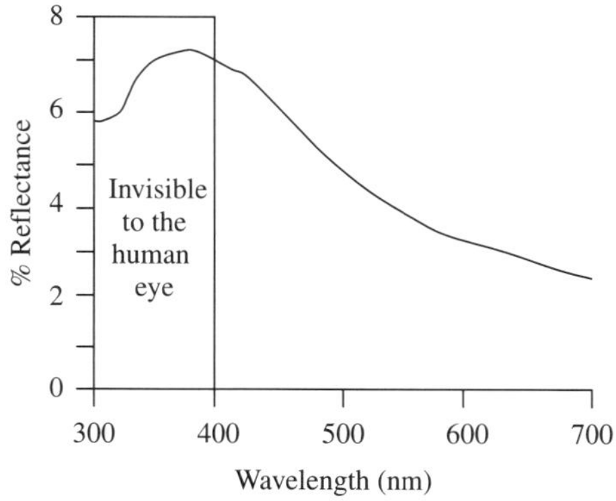

Figure 7.11 Mean reflectance spectrum for contour plumage of male satin bowerbirds. The plumage reflects most light in wavelengths invisible to the human eye. From Doucet and Montgomerie (2003).

run counter to the notion that rate of daily growth of an original feather can be used to predict male mate quality the following breeding season.

7.7 Fluctuating asymmetry

The final works to be reviewed in this section use rate of feather growth as a standard against which to assess variation in fluctuating asymmetry, another trait that has been considered an honest signal of mate quality. First, some background is necessary on the concept of fluctuating asymmetry and its presumed relation with nutritional condition (Leary and Allendorf 1989; Parsons 1990). Bilaterally symmetrical animals consist of halves that are mirror images of each other. It is thought that one set of genes controls development of both sides, giving the mirror-image effect known as bilateral symmetry. Nevertheless, many bilateral structures have been found to be asymmetrical, with some degree of variation existing between the left and right sides. Because there is no particular reason for a trait to have a larger value on one side or the other of the body, the symmetry is said to be fluctuating. (Certain traits, such as claw size in fiddler crabs, are exceptions.) Both genetic and environmental factors have been shown to affect degree of asymmetry. For example, inbreeding leading to high levels of homozygosity has been associated with increased fluctuating asymmetry (Leary and Allendorf 1989). In a mating system context, it has been hypothesized that the extent of asymmetry in bilateral structures can be used to assess quality, that is, resistance to nutritional stress, during the growth of those structures, and so provides an honest signal of quality for mate choice (e.g. Møller 1992).

Møller (1992) had found that the degree of symmetry in the length of the two outmost tail feathers of male barn swallows was used in mate choice by females of the species. In a follow-up study (Møller 1996), he employed ptilochronology to determine whether the asymmetry signal was an honest indicator of male quality, that is, whether males in better nutritional condition actually did grow L6 and R6 rectrices that were more symmetrical. While he was at it, he examined the relationship in females as well. The swallow population he studied breeds in Denmark and winters in Africa. Møller assumed that environmental conditions during the birds' normal winter molting period in Africa were less stressful than they were while the birds were raising young in Denmark. He predicted that growth bars on the two outermost rectrices would be wider and more symmetrical on the original feathers grown during the normal winter molting period than on induced rectrices grown during the breeding season.

Original rectrices were plucked early in the breeding season. The birds were recaptured and induced rectrices pulled at the end of the season in July and August. Degree of asymmetry was calculated as the absolute value of the difference in various measures between the left and right outermost rectrices.

There was very good agreement between growth bar width and feather length on the one hand and fluctuating asymmetry on the other. Growth bars of both males and females were wider and feathers longer overall on original than on induced rectrices. In parallel, bilateral symmetry of both growth bars and feather lengths was greater in original than in induced rectrices.

Møller also found a positive correlation between the degrees of fluctuating asymmetry of original and induced feathers. That is, the more symmetrical were feathers produced during the winter molt, the more symmetrical were the induced rectrices produced by the same bird during the breeding season. This finding is important for it buttresses the assumption of earlier papers in this chapter that characteristics of original feathers grown during the previous molt can carry over as indicators of quality during the following breeding season. While Møller's project suffered from the uncertainties associated with correlational evidence, the next study tied feather growth characteristics to feather asymmetry in a manipulative, controlled experiment.

As a complement to his study of food-hoarding in Eurasian nuthatches discussed earlier, Jan-Åke Nilsson used ptilochronology to examine whether nutritional condition might be related to the fluctuating asymmetry of tail feathers in nuthatches (Nilsson 1994). He imagined that even though length of tail feathers is not likely to be subject to sexual selection in nuthatches as it is in barn swallows (Møller 1992), the extent of the feathers' asymmetry could provide information about quality to other nuthatches, both female and male.

Recall that in their study of food caching, Nilsson and colleagues (Nilsson *et al.* 1993) provided extra food for one month to nuthatches in the treatment group to cache within small woodlots. The birds in the control group were unsupplemented. Between a month and a half and two months later, birds in both groups were captured, had their fourth rectrices (L4 and R4) plucked, and were released back into their territories. Finally, in March, the researchers collected the pair of induced feathers from each nuthatch.

From the hypothesis that fluctuating asymmetry is a positive function of nutritional stress (or a negative function of nutritional condition), Nilsson predicted that the original pair of rectrices grown during the normal molting period should be least asymmetrical, the induced rectrices of birds with access to cached supplemented food intermediate, and the induced feathers grown by birds without access to supplementary food most asymmetrical.

For this analysis, Nilsson compared only feather length with fluctuating asymmetry, but cited earlier work (Nilsson *et al.* 1993) that induced growth bar width and feather mass were also greater in the supplemented birds. The average length of the L4 and R4 rectrices was longest for the original, intermediate for the winter-supplemented induced, and shortest for the winter-unsupplemented induced feathers. From these records, Nilsson concluded that nutritional stress during feather growth varied as original < winter-supplemented induced < winter-unsupplemented induced. He then checked to see if degree of feather fluctuating asymmetry would follow the same pattern. It did; fluctuating asymmetry was least in original feathers and most in unsupplemented induced rectrices. These results led Nilsson to conclude that the rate at which net energy is secured can be a factor influencing degree of asymmetry. In the case of nuthatches, the increase in fluctuating asymmetry could not only have an effect on mate choice, but also result in lower flight performance and reduced ability to evade predators.

7.8 *Summary*

The nine projects in this section on individual quality have taught us much about the possible use of ptilochronology in the context of mate choice and male-male competition. Despite my early emphasis on using induced feathers to control for time and place of feather growth among individuals, six of the nine projects relied on growth of original feathers. Thus, there was certainly variation in the timing and location of the feathers providing indices of feather growth rate. Of these, Takaki *et al.*'s (2001) project relied solely on rectrix growth but, as far as I can tell, of the remaining projects that compared feather growth with plumage characteritiscs, only in Doucet's and Montgomerie's study of satin bowerbirds and Senar *et al.*'s project with great tits were the tail feathers providing indices of nutritional condition and the contour (body) feathers providing scores of coloration grown during the same molting period. In all the other cases, the tail was molted in the autumn and the contour feathers in late winter. Thus, it is entirely possible that the nutritional state of a bird had changed between the two molts, decreasing the tightness of the relationship between apparent nutritional condition and ornament characteristic. The importance is apparent of using induced feathers wherever possible (Nilsson 1994; Senar *et al.* 2003) or, at least, of recognizing the danger of probing for correlations among events happening at different times in birds' annual cycle.

A lesson from these projects is that we cannot assume, as I did early on (Grubb 1989, 1991), that all birds in a study are under the same (mild) nutritional circumstances during the normal autumn molt so that variation between individuals in the growth rate of original feathers is nil. In fact, several of the projects in this section relied on the assumption that growth of original feathers would not be equal, but would be variable due to variable nutritional condition. I no longer advocate using the ratio of induced to original feather growth rate to standardize for birds' structural size, or even to use growth rate of the original feather as a covariate in analyses of induced feather growth. The best method available now for correcting for bird structural size would seem to involve use of some aspect of skeletal structural size, such as tarsus length, as a covariate in multivariate analyses. Of course, even better are manipulative studies where birds are randomly assigned to treatment groups to control for variation in structural size.

Out of this section has come important evidence linking more closely nutritional condition indexed by feather growth rates to traditional measures of relative fitness. Takaki *et al.* (2001) linked feather growth rate positively to both survivorship and reproductive success. Keyser and Hill (2000) positively related the feather growth index of nutritional condition to territory size, body size, and nestling-feeding rate.

Finally, Nilsson supported the possible signal function of fluctuating asymmetry by showing, in a controlled study of feather growth, that degree of fluctuating asymmetry is inversely related to nutritional condition at the time the asymmetrical structures are being developed.

8 *Reproductive effort*

In this chapter, we review eight reports concerning a variety of ways that ptilochronology has led to new insights about a conceptual area of reproductive biology termed reproductive effort. Sometimes called parental effort, reproductive effort was defined by Trivers (1972) as "any investment by a parent in an individual offspring that increases that offspring's chance of surviving at the cost of the parent's ability to invest in other offspring." Inherent in this definition is the notion of trade-offs to be made by parent birds. For example, does a parent invest its efforts equally in all this year's offspring or does it favor only a few? A number of raptorial birds seem to take the latter course. At least, they do not make an effort to distribute food equally among offspring. The result is that while in a year of abundant food all nestlings survive to fledging, in a lean year the first-hatched young use their temporal advantage to take a disproportionate share of the food brought by the parents, leaving their later-hatched, smaller siblings to die of starvation. A second trade-off, more difficult to detect, is one in which the extent of a parent's reproductive effort in one year is traded off against such effort in succeeding years. The argument goes that reproductive activities have a cost in a decreased chance of surviving until the next breeding season. Thus, there is a trade-off between engaging in more reproductive effort this year and, perhaps, raising more offspring this year, or engaging in less reproductive effort this year and having a higher probability of surviving until the next and succeeding breeding seasons (Bryant 1979; Nur 1984). Rate of feather growth has been used in this context as an index of reproductive effort, an index that points to consequences of effort too small to be detectable by changes in survivorship. As we will see, some of this work has tested the notion that mean annual survivorship should influence extent of parental effort. Birds of long-lived species should invest less in any given year's reproduction than should birds of short-lived species.

8.1 *Brood size*

We start with research showing that, in fact, there is a cost to reproduction in lost nutritional condition. The aim of an experiment by Douglas White, Dale Kennedy, and Philip Stouffer was to test a central idea of parental investment theory, namely, that raising more young is more costly to parents than raising fewer young (White *et al.* 1991). If such an idea is true, they reasoned, European starlings raising more offspring should be in poorer nutritional condition than those raising fewer offspring. The indices of nutritional condition were the rate of daily growth as indicated by growth bar width (obtained with the usual pin and card method), feather total length, and feather mass of an induced feather being grown while birds were bringing food to their nestlings. They tested the prediction in a well-designed experiment using starlings nesting in boxes. Although their experiment included some females that laid second clutches, confining our attention to first broods satisfies the objective of showing that reproduction comes at a cost in nutritional condition.

The use of nest boxes permits close monitoring of breeding events, allowing for experimental precision. White and co-workers plucked the outermost left rectrix of female starlings captured in nest boxes on the sixth day after their eggs had hatched. Before being released back into their nest boxes, the birds were banded, weighed, and had their wing lengths measured. The workers knew that in their study population, the natural brood size varied from two to six nestlings with 86% of all broods having between three and five young. On the same day that their mothers were plucked, broods of two, four, or six were formed by moving same-aged young between nests. When the nestlings were between 12 and 17 days old, a period during which reproductive effort of parents was known to be constant, the work rate of all females was measured by counting their number of feeding visits to their young for 10 minutes during one day. In a nice design feature, these feeding counts were alternated among nests in the three treatment groups. The researchers obtained the aggregate weight of each brood on day 17, about the age when nestlings obtain their maximum weight.

This work was done before we realized that the assumption was not universally correct that growth of original feathers was equal among birds of similar structural size. Therefore, growth of the original feather was used as a statistical covariate to control for bird structural size.

We might take a moment here to explore the implications of what we now know was an imperfect idea. Suppose birds that grow an induced feather faster because they are in better nutritional condition at the time also grew their original feather faster during the normal molt because they were in better nutritional condition then as well. If we assume that all birds grow original feathers equally fast, we actually bias results against the fast induced feather growers because the difference between original and induced is greater than we assume. It is only in the case where one bird is in relatively better nutritional condition when growing the original feather and relatively worse condition when growing the induced feather, that we bias toward finding a difference between the two birds' induced feathers. While this second event is unlikely to happen often, it is possible. We now know it is better to use tarsus length or some other measure of skeletal size as a covariate in multivariate models.

In the starling study, no difference was found among original feathers of females in the three treatment groups, suggesting that in this highly controlled study, there was no confounding effect of variation in growth of original feathers. The frequency of feeding visits by the female increased significantly

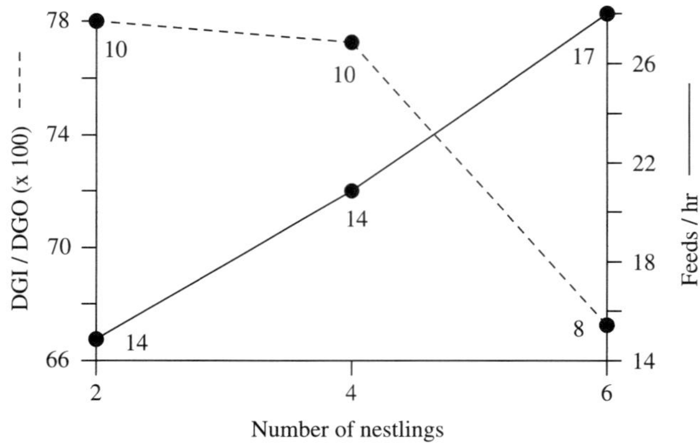

Figure 8.1 Increasing the brood size of European starlings from two to six caused an increase in the number of feeds of the brood per hour by the female parent and a decrease in the rate at which she grew an induced feather while doing so. Sample sizes are shown in the figure. From White *et al.* (1991).

with brood size, 15, 23, and 28 feeding visits per hour, respectively, for broods of 2, 4, and 6 (Figure 8.1). Thus, females with larger broods worked harder.

Feather growth indicated that females feeding broods of six were in relatively poor nutritional condition compared with those feeding two or four nestlings (Figure 8.1). Daily induced feather growth of females feeding six young was only 69% of daily growth of the original rectrix; analogous percentages were 78% and 77%, respectively, for birds feeding two and four young. Mass of the induced feather decreased with increasing brood size, but there was no difference among lengths of induced feathers in the three groups.

The contrast between results for growth bar width and total length raises an interesting matter. Even though the six-nestling females grew their feather considerably more slowly per day, they grew it to the same total length. The only way this could have happened is if they grew it for more days. Does this imply that the follicle is somehow monitoring daily growth and making suitable adjustments to growth period? Does it mean that the follicle has dumped on its doorstep, so to speak, a feather's worth of material and stops growing the feather when the allotment of material is used up? In general, how does a follicle know when to stop growing a feather? Nobody seems to have investigated these questions.

That feeding rates and feather growth records were about the same for broods of two and four brought forth from the authors a possible explanation involving impact of greater thermal costs on work rate of parents. When huddled together, broods of two would have a higher surface to volume ratio than broods of four. Thus, heat loss per nestling would be higher in broods of two. The increase in thermal cost per nestling in groups of two would slow growth rates unless parents worked harder per nestling, hence the reason that feeding rates and feather growth rates were similar in groups of two and four.

White and colleagues interpreted their results to mean that nutritional condition declined non-linearly with increasing brood size. Recall that all the brood sizes used in the study had been found naturally, implying that some starling females, those with large broods, incur a nutritional cost for their relatively great reproductive workload. The increased reproductive effort did have a cost. It would have been quite interesting to know something about survival rates so we could judge whether females feeding broods of six had been forced to sacrifice future reproductive success by incurring a higher annual mortality rate. Such a difference might well be small enough to require many years of experimental replication to detect.

Carrying forward from the starling study the idea that breeding has a cost in lost nutritional condition, we now embark on a series of reports exploring how such nutritional costs may have affected the life-history strategies of birds. To an evolutionary biologist, the term, strategy, does not mean that an animal is carrying out some conscious plan aimed at attaining a particular goal, the common application of the term to human behavior. Rather, to an evolutionary biologist, strategy means a series of linked behavioral patterns that have been favored by natural selection because they confer increased relative fitness. For example, we do not imagine that a bird has done the calculation and has decided that laying four eggs per clutch will give it the greatest lifetime reproductive success. We do imagine that over evolutionary time, birds of that species, age and experience, and in that particular habitat, which lay four eggs per clutch leave more genes per lifetime than birds that lay a greater of lesser number. This logic holds regardless of the bird's state of consciousness.

8.2 *Breeding and molting*

The next three papers all concern descriptive studies that use ptilochronology to help explore the interrelation of reproduction and molting, life-history stages that are both impacted by nutritional condition. In many breeding populations of songbirds, some pairs attempt to raise two broods of offspring in a breeding season while other pairs attempt to raise only one brood. Such is the case with Bridget Stutchbury's study population of hooded warblers in northwestern Pennsylvania, so her student, Leslie Ogden, and she attempted to understand the causes of the variation (Ogden and Stutchbury 1996). Over the course of a 5-year project, they discovered that double-brooded pairs raised an average of 1.9 more offspring per year than did single-brooded pairs. The extra young amounted to an increase of 63% per year in reproductive success for double-brooded pairs. Given such a huge difference in annual reproductive success, at least of fledglings, the researchers set about finding a cause for why all pairs were not double brooded. Truth be told, their very extensive data analysis came up empty as to evolutionary cause. They were not able to show that double-brooded birds had reduced survivorship until the next breeding season compared with single-brooded birds. However, some of their findings suggested the mechanism for single- or double-broodedness and that a difference in survivorship might have emerged with larger sample sizes.

The first clue to the mechanism was that birds that started their first breeding attempt early were more likely to attempt to breed twice. So it seemed likely that limited time might constrain double-brooding. As do many songbirds, hooded warblers molt at the end of the breeding season, completing a full molt of body and flight feathers before departing on migration. Ogden and Stutchbury found that double-brooded parents tended to still be feeding second-brood fledglings while molting, and completed their molt a full three weeks later than single-brooded birds. Ptilochronology was employed to investigate whether the overlapping of parental care and molting exacted a nutritional cost from double-brooded parents.

During the molting period and just after, birds were driven into mist nets by teams of "beaters," and a newly molted rectrix collected. Growth bar width was measured with a caliper. The difference between growth bar width of a rectrix molted by a double-brooded parent while it was still feeding second-brood fledglings and a rectrix molted three weeks earlier by a single-brooded parent was not significant, so there did not appear to be any cost in nutritional condition to raising two broods. It would be interesting to know if, regardless of broodedness in one year, birds with wider growth bars on original rectrices molted in the fall were first to arrive on the breeding ground the following spring. Such a result would indicate some inherent difference in quality similar to that found in Styan's grasshopper warbler by Takaki *et al.* (2001).

After an extensive analysis had failed to turn up any other potential mechanism, Ogden and Stutchbury focused on the consequences of the three-week difference in molt timing as a possibly important explanation. Hooded warblers are known to defend feeding territories on their wintering grounds. Early-arriving birds on those wintering grounds presumably secure the better (e.g. food-richer) territories with very late-arriving birds not achieving a territory at all, but remaining as non-territorial "floaters" (Stutchbury 1994). Thus, there could be a big cost to double-brooding and late molting if the resulting delay in migration results in a lower food supply on the wintering ground. Even though not shown by the small sample sizes in this study, we might expect to see a relationship between departure date on fall migration and survivorship until the next breeding season. Ogden and Stutchbury also advance the very nice prediction that since migratory species without winter territoriality could well suffer a lower penalty for arriving late on the wintering ground, such species should have a higher frequency of double brooding.

We know from our analysis of cardinal feathers (Grubb *et al.* 1991), that feather growth rates are greatest during the normal autumn (i.e. pre-basic) molting period. This hooded warbler study seems to indicate that in that species, nutritional constraints may not be very important at that time. Feeding young at the same time had no effect on rate of feather growth. So, the constraint may be confined to the physiological level, to the rate at which a follicle with ad libitum energy and nutrients can incorporate that energy and those nutrients into a growing feather. Perhaps, because of constraints at the molecular level, the rate of feather production during the normal molting period has somehow reached a point where it can no longer be accelerated by natural selection. The next paper takes this point of view in exploring the mutual constraints of breeding and molting in albatrosses.

Laysan and black-footed albatross nesting colonies are found on several of the leeward Hawaiian Islands. While the birds have generally been thought to breed every year, two long-term studies of banded birds found that each year about 25% of adults did not breed and apparently did not even return to the breeding island (Rice and Kenyon 1962; Fisher 1976). The vast majority of such non-breeders were still alive because they resumed breeding at the same location the next year. A second salient fact is that birds of these two species do not molt all of their primary flight feathers every year during the non-breeding season as most bird species do. Instead, they often arrive to begin breeding while still carrying on each wing one or more worn primaries that have been left unmolted. The birds do molt their outer primaries every year, those that receive the most wear, but often not all the inner primaries are molted.

Nancy Langston and Sievert Rohwer (1996) combined these two facts to produce a new hypothesis, that since breeding annually and molting a complete set of primary feathers annually would take more than 365 days, natural selection has favored a compromise between reproduction and molting. Results from ptilochronology were pivotal in supporting this idea empirically.

These two species of albatrosses forage for food over vast expanses of the Pacific Ocean. While a good deal had been known about their breeding biology on the island, only recently have birds that drowned in driftnets on the high seas during other times of the year become available for study (Johnson *et al.* 1992). The researchers knew that Laysan and black-footed albatrosses, respectively, spent about 260 and 245 days a year engaged in breeding activity. Since none of the breeding birds was also growing feathers, it was known that breeding and molting periods did not overlap. Langston and Rohwer (1996) had the idea that albatrosses of these two species did not have enough time in any one year both to produce an offspring and to molt a complete set of new primaries. They reasoned that natural selection had favored a compromise, the birds did not grow all their primaries most years, and every four years, or so, they skipped breeding so that replacements for the worn primaries that had accumulated would have time to grow in over a longer time period during a complete molt. They knew how many days were devoted exclusively to breeding, leaving Laysan and black-footed albatrosses, respectively 105 and 120 days to molt in the absence of breeding obligations. What they needed to know was how fast the birds grew their primaries so they could calculate about how many days a year the molting of primaries would take.

They obtained the feather growth rate information by measuring growth bars on the primaries of birds killed in nets. It turned out that birds replacing only the outer primaries, those that are replaced every year, grew each feather at the rate of 5.42 mm per day. A very important fact emerged when they calculated the daily growth rate of primaries of birds that were growing in both outer and inner primaries simultaneously; such birds grew the increased number of feathers at the rate of 5.44 mm per day, essentially the identical rate as for singletons.

An aside is required here on how daily growth was calculated. The researchers first used the established method of considering one growth bar (i.e. one combination of a light band and a dark band) to represent 24 hours growth of the feather. However, using this value resulted in an impossibly long time, 262 days, to grow in all 10 primaries. Langston and Rohwer knew that these albatrosses eat mostly squid, a prey item available to the birds only at dawn and dusk when the squid are at the surface and there is enough light for the birds to see them. Although they did not know the precise mechanism producing growth bars (no one does), the authors reasoned that if there are two active feeding periods and two rest periods per 24 hours, then perhaps two growth bars are created per 24 hours. They operated on this assumption to arrive at the figures for daily growth mentioned earlier. (A study of reproductive effort in Leach's storm-petrel that we will encounter shortly had similar problems interpreting growth bars. Storm-petrels are close relatives of albatrosses, so the difficulty in interpreting growth bars may have similar causes in the two species.)

Because individual primaries growing alone grew no faster than when growing with other growing primaries on the same wing, Langston and Rohwer concluded that while the birds were on the high seas during the non-breeding season, there was essentially no energy or nutrient constraint of an ecological nature on rate of feather growth. That is, the birds could find all the food they needed to molt many feathers at once. Instead, they suggested, there was a constraint of a physiological nature, that a feather follicle was at the rate limit of how fast it could assemble energy and nutrients and incorporate them into a growing feather. According to the researchers' calculation, a bird replacing all 10 primary feathers on a wing would need to grow a total of 2345 mm of feathers. Dividing this by 5.4 mm, the rate of daily feather growth, would give the number of days that a complete molt of the primaries would take if a bird grew one primary at a time. However, the birds grew an average of 3.3 feathers at once, so the researchers also divided 2345 mm by 3.3 to account for the simultaneous growth. The result of this double division was an estimate of 131 days needed to grow a full set of primaries. Since only 105 to 120 days were available between breeding seasons, Langston and Rohwer argued that natural selection has favored growing less than a complete set of primaries every year. The result is that old, worn feathers accumulate to the point where skipping a breeding season has been selected for so the birds have time to grow a complete set of new primaries.

Through study of how molting was progressing on the birds killed at sea, the researchers determined that about 3.3 new primaries per wing were being grown simultaneously. One way that natural selection could have gotten around the follicle-level growth constraint would be to grow in more feathers at the same time. Langston and Rohwer argued that growing more feathers at the same time has to mean losing more old feathers at the same time and such a strategy would be selected against because more and/or larger gaps in a row of primaries would hinder the birds' flying performance, thereby increasing the energetic cost of flight and raising their risk of death in storms.

Langston and Rohwer point out that this albatross strategy is to be expected in such birds that have a very long life expectancy, on the order of decades. Such a strategy tends to sacrifice reproductive effort in the present year if doing so increases the chances of many more years of reproduction. Short-lived birds should have the reverse strategy, large reproductive effort in the present year because the probability of living into the future is low. For example, among the songbirds with their annual mortality rate of about 50%, it is not uncommon for late-molting adults essentially to lose the power of flight temporarily because they are molting so many feathers at the same time (Jenni and Winkler 1994).

Langston and Rohwer concluded their report by advancing several predictions from their breeding/molting trade-off hypothesis to be tested with future work. First, they predicted that birds that fail in their breeding attempt one year should be able to start molting early and, therefore, should be more likely than successful breeders to return to breed the next year. In fact, it might be added that nearly 100% of birds with breeding attempts failing early, say from an egg failing to hatch, should attempt to breed the next year. Second, if one caught birds at the breeding colony and examined the state of their molt, birds with a greater number of worn primary feathers should be less likely to breed the next year because of the requirement to undergo a complete molt of the primaries.

Albatrosses mate for life and it is known from the work of Rice and Kenyon (1962) and Fisher (1976) that if one member of a pair fails to return to the breeding colony, presumably having died, it takes the remaining bird of the pair two years to find another mate and begin breeding again. That is, there is a year of reproduction lost in the process of obtaining a new mate. Langston and Rohwer's third and most intriguing prediction stems from this fact. The prediction is that the two birds in mated pairs should arrange things so that they both refrain from breeding in the same year in favor of both molting all primaries during the interval. If this were not so, if one bird returned to breed and the other stayed away to molt, the returning member of the pair would be unsure whether its mate had died and that it should begin searching for a new mate.

How might such synchronization of complete molts work? What might be the so-called proximate mechanism? A proximate mechanism refers to the environmental cause in the here-and-now, not the evolutionary cause. Evolutionary causes are sometimes called ultimate causes because they are explained ultimately in terms of evolutionary fitness. In this case, we can understand what the ultimate cause might be, a bird that skipped breeding in the same years that its mate did would leave more offspring during its life than one that did not. But what might the proximate mechanism be? A couple of possibilities come to mind. First, suppose that during its first breeding season, a bird becomes mated to another bird also in its first breeding season, and suppose the year-to-year accumulation of worn primaries is fairly constant from bird to bird, then the two birds would be naturally synchronized in the timing of their years off from breeding. It is also possible that each bird is aware of the number of worn primaries carried by its mate and can remember such information from the end of one breeding season to the beginning of the next. If a mate known to have had many worn primaries the previous year did not appear at the beginning of a breeding season, the returning mate could attribute the absence to a molting requirement, and abandon that year's breeding itself to molt all its own primaries. In this regard, the spread-wing postures of albatrosses during courtship displays may present each courting bird with information about the worn primaries on its partner.

Before leaving this albatross story, we should remind ourselves that predictions like these should be examined with manipulative tests rather than just the descriptive work mentioned so far. A crucial part of manipulative tests is the random assignment of birds to treatment and control groups, a procedure that has the powerful feature of isolating the variable being examined from other potentially correlated variables. For example, we might predict that randomly selected pairs of albatrosses from which we steal the egg soon after it is laid would be much more likely to return the next year (with a full complement of new primaries) than control pairs.

Albatrosses, shearwaters, and storm-petrels belong to the avian Order Procellariiformes, the so-called tube-nosed swimmers. Cory's shearwater is a medium-sized bird of this order, about the size of a ring-necked pheasant, which breeds on islands in the Mediterranean and eastern Atlantic. In studying the molt pattern of this species, Montiero and Furness (1996) pulled two primaries and one rectrix from each of several birds so they could get a reading on daily feather growth, as in the albatross study. The result of this effort was quite instructive.

Cory's shearwaters require 207 days to molt all their primary feathers, a considerably shorter period than the 260 days for Laysan and 245 days for black-footed albatrosses. One reason for the difference is that because shearwaters are smaller than albatrosses, each feather is shorter. For this reason, shearwater feathers may take less time to grow fully.

Monteiro and Furness pulled feathers to find out how fast an individual shearwater primary grew per day, but were met with a surprising result. Unlike the albatrosses, Cory's shearwater molts during the breeding season, at least the body feathers. While the molt of primaries seems to conclude soon after the breeding season commences, the rectrix molt starts before the breeding season begins, is suspended during reproductive activities, then finishes with the molting of the last few rectrices after the breeding season concludes. The researchers pulled primaries and a rectrix at the start of the breeding season, assuming that induced feathers would be grown at that time since new body feathers were appearing. Having pulled original feathers in mid-June, the workers caught the birds again no earlier than August, but discovered that none of the five study birds had started to regrow a single feather. I mention this result because it reinforces our early findings with the annual molt of cardinals (Grubb *et al.* 1991). As Montiero and Furness conclude, the suspension of tail molt and the lack of induced feather growth during June and July, a time when body feathers are being vigorously molted, indicates that molt in this species is strongly influenced by intrinsic physiological rhythms. The only similar case we have encountered so far concerned lack of growth of induced feathers by unsupplemented Alaskan gray jays in Tom Waite's caching experiment. The lack of regeneration there, however, seemed clearly a function of inadequate nutrition because the birds that had been allowed to cache supplemental food all grew out induced feathers. If some way could be found to supplement the food source of these shearwaters, might they grow induced feathers right though the breeding season?

We conclude this section on the timing of feather growth with an intriguing and possibly unique phenomenon, the growth of the two central rectrices of red-tailed tropicbirds. These beautiful creatures breed on tropical and subtropical oceanic islands; those breeding on French Frigate Shoals Atoll in the northwestern Hawaiian Islands have been studied for a number of years. While brilliant white in overall plumage, each tropicbird possesses two central rectrices that differ from its other rectrices in being both very much longer and blood red in color rather than white. Termed tail streamer ornaments, these two much-elongated feathers are twitched from side to side in a visual display during the aerial courtship maneuvers of pairs. Using measurements of growing streamer ornaments taken on the bird at intervals over the course of the breeding season (rather than measurements of growth bar widths), Allison Veit and Ian Jones (2004) confirmed that, growing at about 2.2 mm/day, each streamer took about 181 days to grow fully and that birds possessed two fully grown streamers at the beginning of their breeding period, just the interval when they are used in courtship. The unusual feature of the system, however, is that the two streamers were molted and re-grown alternately rather than simultaneously. This is the only example I know of in which, under natural conditions, two putatively bilaterally symmetrical feathers grow so far out of phase. One possible explanation for this asymmetrical growth, offered by Veit and Jones, stems from the fact that the species inhabits low latitudes. On islands near the equator, individual pairs of birds are known to begin breeding at any month of the year. If a bird paid attention to the relative lengths of potential mates' tail streamers and courted only birds whose streamers were of equal length, it would be confining its attention to birds coming into breeding condition regardless of time of year. The proximate mechanism responsible for having just one tail feather, of the many thousands of feathers on a tropicbird, molted and replaced outside of the normal molting period seems to be unknown.

8.3 *Tests of theories of reproductive effort*

We now come to three projects that employed ptilochronology as part of manipulative tests of theories of reproductive effort. The first two studies concern the trade-off made by parents between present and future reproductive effort, while the third report deals with how production of sons or daughters can be viewed in the context of reproductive effort.

Leach's storm-petrel, a bird the size of an American robin or European blackbird, is another member of Order Procellariiformes, an avian order known for the length of individual life-spans. Many storm-petrels are known from long-term banding studies to be more than 20 years old, while a 5-year-old robin is quite an old bird. Harking back to the early part of this chapter, we remember that the theory of reproductive effort predicts that animals of long-lived species, with their greater chances of reproduction in the future, should put less effort into present reproduction than should individuals of short-lived species. Because Leach's storm-petrel is a long-lived species, Bob Mauck and I (Mauck and Grubb 1995) predicted that if we handicapped birds to artificially increase their workload while they fed their nestling, they would shunt the cost of the extra work to the nestling rather than bear it themselves.

Our study site, Kent Island, New Brunswick, Canada, is an isolated partially conifer-covered islet at the southern end of the Grand Manan Archipelago in the Bay of Fundy. There, the storm-petrels nest by the thousands in underground burrows which they dig in the soft peat. The parents bring food to the one nestling in the burrow from distances of, probably, several hundred kilometers out to sea. During the nestling-feeding period, each parent commutes with food to the nest about every 48 hours, coming and going only at night.

The handicapping treatment was applied to both members of the 10 treatment pairs. After catching parent petrels in their burrows, Mauck and I increased their work rate by clipping 1.5 centimeters from the end of each wing. According to the aerodynamics equations of Pennycuick (1989), the total wing span reduction of 3 centimeters increased the petrels' cost of flight and, therefore, work rate during feeding flights, by about 9%. The restriction on the flight performance of parents was only temporary as the birds would grow a new set of full-sized primaries during the post-breeding molt. The 20 birds in the 10 control pairs were handled in an identical manner with the exception that we merely touched the scissors to the wings. After plucking the original R6 rectrix, we allowed treatment and control birds to return to their burrows. To control for possible variation in responses due to hatch date, assignment to treatment or control group was decided by coin toss within successive pairs of burrows based on hatch day.

Several weeks later, we recaptured adult petrels by trapping them in their burrows as they returned to feed the chick. We checked the traps at about 04:00 each night and when we found an adult, we removed it from the burrow, pulled its induced R6 rectrix, and then released it back into its burrow.

If the parental care exemplified by carrying food great distances to the chick were to become unnaturally costly due to a handicap such as the one we imposed, a petrel might reduce its own nutritional condition, the nutritional condition of its nestling, or both. We used growth rates of the induced R6 rectrix to monitor the nutritional condition of parents and continually monitored body growth rate as an indicator of the nutritional condition of the nestling. As Leach's storm-petrel is a relatively long-lived species, we predicted that birds with clipped wings and, thus, harder work-rates would grow a tail feather as rapidly as would controls, but the nestlings of the wing-clipped adults would grow less rapidly. That is, the adults would sacrifice present reproductive effort in the interests of preserving lifetime reproductive success. They would shunt the nutritional cost of the extra work rate to the current year's offspring.

Langston and Rohwer (1996) concluded that the albatrosses they studied produced two growth bars per 24 hours. The induced feathers on the Leach's storm-petrels that Mauck and I studied did not exhibit any regular array of growth bars. Instead, there was such a puzzling inconsistency to the light and dark bands that we were not able to extract any measure of growth bar width. It may be that we were dealing with a similar phenomenon as that shown by the albatrosses, except that instead of two regular periods of active feeding per day, our petrels displayed a complex irregularity of feeding intervals, perhaps due to an interaction between day/night photoperiods and the strong tidal upwellings in the area. In any case, we used as a measure of nutritional condition of the parents the total length of the induced feather divided by the total length of the original feather divided by the number of days between plucking the original and induced rectrices. This procedure gave us the proportion of the original feather that the induced feather grew per day. In cases where we retrieved induced rectrices from both parents of a particular nestling, we used in our analysis the average of their feather growth indices.

Nestling growth was monitored daily between days 7 and 44 post-hatch (during the daytime when the parents were not in the burrow). From these daily measures of nestling weights, we calculated the proportion of all nights during that period during which the nestling was fed by at least one parent.

Over the course of the experiment, control nestlings grew at a mean rate of 1.39 grams per day, while treatment nestlings grew at a mean rate of only 1.12 grams per day, a statistically significant difference (Figure 8.2a). In line with this difference in growth rate, we found that, on average, control nestlings and treatment nestlings were fed on 70% and 64% of nights, respectively, again, a statistically significant difference.

Twenty-nine of the 40 adults in the study were recaptured and their induced feathers collected. We found no difference in the index of feather growth rate between treatment and control groups. While the control adults grew their

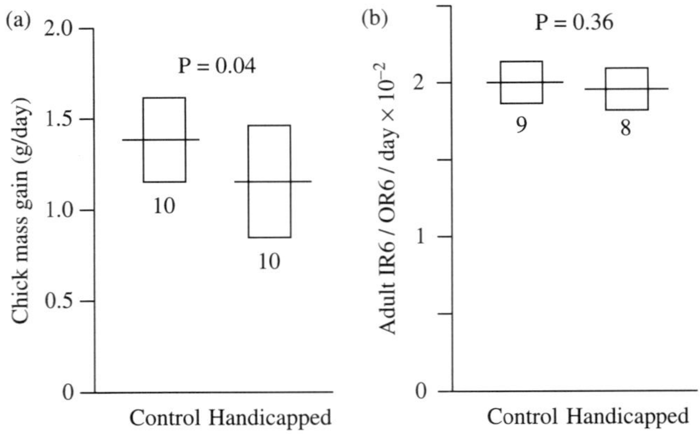

Figure 8.2 Mean ± SD (a) chick daily mass gain and (b) adult feather growth index for control and handicapped breeding pairs of Leach's storm-petrels. Parents shunted the cost of the handicap to their offspring. Probabilities of significance and sample sizes, respectively, are shown above and below the box plots. From Mauck and Grubb (1995).

induced rectrix an average of 1.99% of the length of the original rectrix per day, the analogous figure for the treatment group was an almost identical 1.97% per day (Figure 8.2).

Mauck and I concluded that the results of this experiment best fit the hypothesis that when faced with increased energetic demands, adult Leach's storm-petrels maintain their own nutritional condition and allocate fewer resources to the nestling. Adult nutritional condition, as indexed by feather growth rate, was not statistically different between treatment and control birds, but treatment nestlings grew at a slower rate than did control nestlings.

While we were not able to monitor directly the number of parental feeding visits to the nestling, the calculated proportions of nights when the chick was fed by at least one parent suggested that the primary method employed by the treatment parents to reduce the cost of the handicap to themselves was to make fewer visits. Treatment nestlings were fed on significantly fewer nights than were control nestlings, suggesting that food delivery was more consistent for control than for treatment nestlings.

This storm-petrel project provided experimental support for the hypothesis that in long-lived birds, parents sacrifice present reproductive effort to preserve lifetime reproductive success. It would be nice to test for a higher subsequent mortality rate in the handicapped group, in line with the theory. Unfortunately, we do not have such records and, in any case, a handicapping rate of only 9% may not be reflected in mortality rates even though it appeared to affect the birds' food-delivery strategy.

It is worth noting how the random assignment of petrel pairs to treatment and control groups alleviated concern about growth rates of original feathers. Mauck and I divided length of the induced rectrix by the length of the original rectrix under the assumption that all birds were equally well-nourished while growing the original feather during the previous molt. However, in the last chapter, we reviewed several studies that revealed considerable variation among individuals in the growth rate of the original feather. By randomly assigning petrels to treatment and control groups, we greatly reduced concern about differences in growth of original feathers because we assumed that, just by chance, equal proportions of more nourished and less nourished birds would wind up in both groups. In short, random assignment is very important because it goes far toward eliminating the possibility that some independent variable other than the one under consideration will bias results in an unknown fashion.

Ian Nisbet, Jeremy Hatch, and colleagues attempted to repeat Mauck's and my handicapping experiment with another long-lived seabird, the common tern (Nisbet *et al.* 2004). While partially confirming our findings, their results raised a number of questions about interpretation of handicapping experiments.

Fourteen male terns from 3 to 21 years of age breeding on a Massachusetts island were handicapped by clipping their wings (Figure 8.3). The shortening was designed to raise the cost of flight by about 10% (Pennycuick 1989). The original right R5 rectrix was collected from these males, from their mates, and from the male and female of unclipped control pairs matched with the treatment pairs for hatch date of the first egg and total clutch size.

Figure 8.3 (a) Jeremy Hatch and Ian Nisbet at work in the common tern colony on Bird Island, Massachusetts. (b and c) Common tern with one clipped and one unclipped wing. (d) Weighing a tern chick. Both wings of each experimental bird were clipped so that the overall wingspan was reduced by 50 mm, increasing the cost of flight by a calculated 10%. Photographs by Jeremy J. Hatch.

At each of two subsequent dates, the growing induced rectrix was measured *in situ* (on the bird) and immediately after the second of these measurements, the induced rectrix was pulled and stored. The researchers claimed that measurements of growth bar widths were less accurate than measurements of total length, so they used the latter to calculate feather growth as an index of parental effort due to wing clipping. For some reason that they do not discuss, in calculating the regression of feather length (mm) on days of growth they use the *in situ* measurements of feather length rather than measurements of the total length of the plucked induced rectrix, measurements they could easily have made on the plucked feathers. It is unfortunate that they did this as it makes more difficult a direct comparison of their results with those of Mauck's and my storm-petrel study. Using the *in situ* measurements allowed them to double their sample size by using two data points per feather (one from each date of measurement) in calculating the regression line of feather growth. To do so was an error because two measurements of the same feather may not be considered to be statistically independent.

To index allocation of resources to nestlings, Nesbit and co-workers calculated for each nestling the slope of the part of its growth curve in which increase in body mass per day was the highest.

The results of this tern study differed substantially from our petrel results. First, the induced rectrix of handicapped males grew significantly more slowly per day than did that of the controls, not at the same rate. This suggests that, unlike our storm-petrels, the male terns accepted some of the artificially increased cost of reproduction themselves. Second, the older nestling of handicapped males grew at the same rate as the older control nestling. This result suggests that handicapped males and their mates allocated to their first offspring the normal amount of resources, that is, they did not withhold allocation of resources to the first nestling as our petrels did to their only nestling. The researchers showed that the second nestling of handicapped males actually grew faster than the second nestling of matched controls, but we should probably not place too much confidence in this result because the treatment and control sample sizes were only three, some of the control males were of unknown age (see page 135), nine eggs of study birds had been broken and replaced with eggs of other birds, and one control pair was matched with two different experimental pairs. If we discount the uncertain result regarding the second nestling, we may interpret these findings as contrary to the notion of delayed reproductive effort in this long-lived tern species. It appears that just the opposite took place; compared with controls, handicapped male parents took on the increased cost themselves (slower feather growth) and passed none on to the nestlings (equal rate of body mass increase). We need more well-designed studies of more species to resolve the question of whether present and future reproductive efforts are influenced by the species-typical life expectancy of the parent bird.

The results of Nisbet and colleagues also concerned age-specific reproductive effort within a species, and while these results do not involve ptilochronology directly, they warrant a digression. Related to reproductive effort is the concept of residual reproductive value (RRV; Roff 1992). The argument goes like this: for any species, as an individual ages, it should, on average, have fewer and fewer breeding seasons left to it to raise young. That is, its RRV should decrease with age. Life-history theory argues that younger breeding individuals, with their high RRV, should allocate resources in favor of their own survival at the expense of the survival of their present offspring. Conversely, old breeders should have low RRV and should, therefore, allocate more resources to their present offspring and less to themselves. Nisbet and colleagues tested this idea by examining the difference with age in the change in body mass during breeding between handicapped and control males. If RRV theory is correct, the difference for young handicapped males should be less than the difference for old handicapped males. That is, because their RRV is higher, young males should not pay as high a price in resource allocation as older males.

Across all treatment and control males, there was no difference in loss of body mass during the study period. However, when the difference was examined in relation to age, just the opposite result emerged from that predicted by RRV theory. Young handicapped males lost more body mass in comparison to similar-aged controls than did old handicapped males. That is, male age was negatively correlated with the difference in the rate of change of body mass in handicapped and control males while they fed nestlings.

There is a way, contradictory to RRV theory, to explain this result and it turns on a difference in male quality. In other work at the same tern colony (Nisbet and Cam 2002), Nisbet and co-workers found that there did not seem to be any decrease in residual reproductive value with age. There was no decline in survival with age and old birds were as successful as younger breeders in raising young. Thus, they argued that in the present experiment the difference with age in weight loss of handicapped birds compared with controls was a simple consequence of a difference in the quality of the birds. High-quality birds, those able to forage well for themselves and their nestlings, survive better than low-quality birds. So, while the comparison for young birds was influenced by the poor response to the treatment by some low-quality males, low-quality males were not found in the comparison of old birds because they had already died. This theoretical conflict between RRV theory and individual-quality theory has not been resolved and is a fertile area in which to pursue research. What we would really like to know is whether old birds were of higher quality when they were young than were other birds that had already died. Such a comparison requires long-term data sets on the order of decades in the case of long-lived species such as procellariiforms and terns.

We have seen how the trade-off between present and future reproductive effort has been considered among species as a function of life expectancy and within a species as a function of age. Alberto Velando used ptilochronology to explore another aspect of life-history theory, the extent to which reproductive effort should be expended differently on sons and daughters (Velando 2002).

The relevant body of theory is called sex allocation theory, a construct that says that mothers should bias the sex ratio of their sons and daughters according to the mothers' nutritional condition and the cost in resources per son or per daughter. Assuming that sons and daughters take different amounts of resources to raise, the theory says that mothers in good nutritional condition should produce more offspring of the more expensive sex while mothers in poor nutritional condition should produce more of the less expensive sex. By following such a strategy, they would maximize their lifetime reproductive success (Trivers and Willard 1973; Frank 1990).

In Velando's bird, the blue-footed booby, there are clearly different costs to raising daughters and sons. Females weigh 31% more than sons, so the assumption is that they require considerably more resources to raise. At hatching, nestlings of the two sexes weigh the same, but female chicks grow significantly faster than males.

Velando had two objectives. First, he wanted to determine whether mother boobies with artificially raised flight costs would preserve their own nutritional condition and allocate fewer resources to their nestlings. He predicted this outcome because boobies are long-lived birds and so should sacrifice present reproductive effort for future reproductive effort. Also, he predicted that such reduced allocation of food to offspring should affect the growth of daughters more than the growth of sons because female nestlings grow faster and reach a higher fledging weight than do male nestlings. He thought this handicapping procedure was a fair test because the amount of food brought to the nestlings by the female parent is more than three times the amount brought by the male parent (Anderson and Ricklefs 1992). Second, he wanted to test the prediction that female parents in better nutritional condition were more likely to lay female eggs, and the reverse for females in poor condition.

Velando conducted the project on one of the guano islands off the Peruvian coast. At 20 nests in the study colony, he captured the female parent at the early nestling stage, randomly assigned 10 birds each to treatment and control groups and clipped the wings of the treatment females to the extent required to increase their flight cost by 5% according to Pennycuick's (1989) equations. Control females were handled to the same extent, but not wing-clipped. Before releasing the females, he plucked the outermost right rectrix. Velando weighed and took the wing length of nestlings when they reached 45 days post-hatch, and caught their mothers and pulled the induced rectrix when the nestlings were 45–50 days old. To index nutritional condition of the breeding females, Velando followed Mauck's and my calculation (Mauck and Grubb 1995) by dividing total length of the induced feather by total length of the original feather and then dividing that result by the number of days between pluck dates to obtain the proportion of the length of the original rectrix grown per day by the induced rectrix. Sex of the nestlings was determined by collecting a blood sample from each bird and then employing the molecular analysis of Griffiths *et al.* (1998).

Due to the vicissitudes of field biology, only nine of the 10 plucked females in each of the control and treatment groups could be caught and re-plucked. During the nestling-feeding stage of the reproductive period, control females grew their induced rectrix at 14.5% of the original feather per day, while the handicapped birds grew out their induced rectrix at 12.2% of the original feather's length per day. The difference between these two rates was not quite statistically significant, but it is quite remarkable when we consider that the handicap of the treatment birds was only 5% (compared with 10% for previous handicapping studies). By contrast, differences in the reductions in body mass and in body mass controlled for structural size of the birds were both statistically significant.

In line with sex allocation theory, the handicapped females allocated fewer resources to their daughters than did the control females, but allocated the same amount of resources as the control females to their sons (Figure 8.4). Daughters of handicapped mothers were 8% lighter in weight at 45 days of age than were the daughters of control mothers, a significant difference, but handicapping had no effect on the growth rate of sons compared to the sons of control females. Furthermore, at 45 days of age, the wings of daughters of handicapped females were significantly shorter than the wings of control daughters, but there was no difference in the wing lengths of sons of treatment and control mothers.

Female blue-footed boobies almost always lay a clutch of two eggs. In a sample of 30 nests, Velando found that, as predicted by sex allocation theory, the proportion of female nestlings was positively correlated with both the body mass and the body condition (body mass relative to structural size) of the female parent when the nestlings were less than 7 days of age. It would have been very interesting to know whether proportion of female eggs was also correlated with growth bar width and/or total length of the original rectrix, but this calculation was not presented.

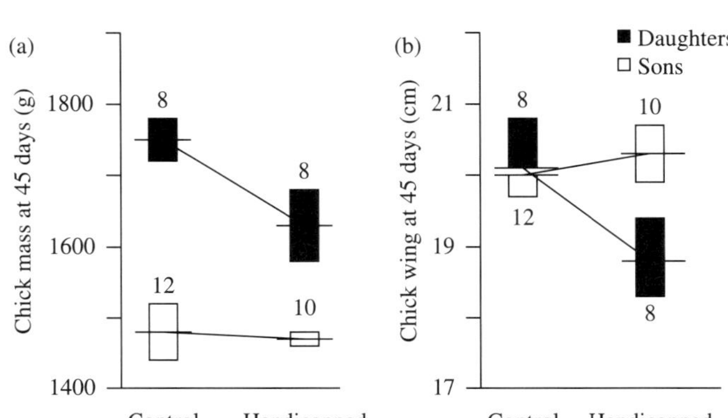

Figure 8.4 Mean ± SE body mass (a) and wing length (b) of 45-day-old male and female blue-footed booby chicks with either handicapped or control mothers. Daughters, but not sons, of handicapped mothers grew more slowly than did control chicks. Sample sizes are shown above or below the box plots. From Velando (2002).

This nicely done project, featuring random assignment of birds to treatment groups, strongly suggests that female parents under more difficult nutritional circumstances favor the growth of sons. What is even more intriguing is the possibility that the nutritional condition of females at the time when they lay their eggs affects the sex of those eggs. This area of life-history study is currently receiving considerable attention. Although we have learned a considerable amount about sperm competition in birds (Birkhead and Møller 1998), the proximate mechanism by which a female might determine the sex of an egg before she lays it is still a mystery.

8.4 *Summary*

The six papers in this chapter have pursued various courses related to avian life-history biology. White and colleagues' experiment with starlings showed a clear trade-off between offspring production and nutritional condition of a parent bird. We await work tying the observed decrease in nutritional condition to decreased lifetime reproductive success. The descriptive studies of warblers and albatrosses and the manipulative study of Cory's shearwaters, taken together, point to the variety of ways that natural selection has resolved the conflict between the energy- and nutrient-demanding activities of reproduction and molting. Finally, the manipulative projects that handicapped storm-petrels, common terns, and boobies leave the impression that more studies will be necessary if patterns are to emerge. It is not at all clear, for example, why the storm-petrel and tern studies produced such dissimilar results.

Finally, that growth bars were difficult to interpret in the albatross (two growth bars per day?) and unreadable in storm-petrels, when combined with another growth bar puzzle to be discussed in the next chapter, indicates that we do not yet have the full answer to the relationship between feather growth and time.

9 *Nestling condition*

I have separated the material in this chapter on the nutritional condition of nestlings from that in the next chapter on prolonged brood-care because they have somewhat divergent theoretical underpinnings. Nevertheless, they can be viewed as a continuum from the day of hatching until the day that juveniles become fully independent of their parents.

In their study of the growth of nestling tree swallows, John McCarty and David Winkler (1999) measured the developing length of a primary feather on the bird, so their results do not use the methods of ptilochronology in a strict sense. However, a portion of their findings is germane to the question raised by Murphy and King (1991) about time lag between a nutritional shortfall and retarded feather growth, and so has been included in this account.

The study was descriptive in nature and probed for factors associated with growth rates of nestling swallows. As part of the project, the length of the growing outermost primary feather was measured on days 10 and 12 after the day of hatching. McCarty and Winkler made the point that the increase in mass of a feather differs from other increases in mass, such as fat content, because feather growth couples increase in mass with the cellular organization and production of a complex structure. They hypothesized that the organizational part of feather growth, involving recruitment and incorporation of material within the follicle, may cause feather growth rate to be less influenced by nutritional condition on the day of growth than during some short-term previous period.

Employing path analysis, a multiple-regression technique used to infer causal relations from a complex of correlations, McCarty and Winkler discovered what appeared to be definite time lags between the state of a causal factor, food supply, and the growth rate of primary feathers (Figure 9.1). Abundance of flying insects of the type known to be brought to nestlings by parents was compared with the nestlings' feather growth rate. The rate of primary growth on 11-day-old nestlings was related significantly to insect abundance on only days seven and nine nestling age. That is, the abundance of insects two and four days before the day feather growth was measured was more likely to have determined the extent of growth than was the abundance of insects on any other day between day seven and day 11, the day of the measurements. These records suggest that it is unlikely that environmental conditions on a given day are the only factor, or even the prime factor, determining feather growth on that day. An immediate conclusion is that in experiments, to increase the chances of detecting differences with treatment, any manipulation should begin several days, at least, before the original feather is pulled. Another conclusion is that we might not expect any sharp variations in daily

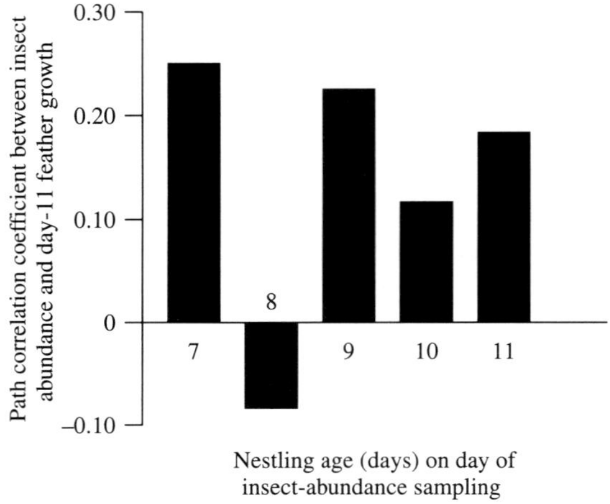

Figure 9.1 Relationships between flying insect abundance on days 7–11 of tree swallow nestling age and the rate of feather growth of 11-day-old nestlings. Path analysis suggested that the extent of the food supply two and four days previously had an important effect on rate of feather growth. From McCarty and Winkler (1999).

increments of feather growth to reflect sharp variations in any causal factor. A blending, an averaging of the effects of any given factor over several previous days, might be expected in feather growth rate on any given day. Thus, experiments that use, say, an A-B-A design of control-treatment-control to examine growth bar widths in different regions of the same feather may become difficult to interpret since the distinction between control and treatment periods in feather growth would be blurred.

The second and last article in this short chapter concerns a puzzle. Michael Kern and Richard Cowie (2002) intended to employ ptilochronology as a tool to study the effect of environmental factors on the nutritional condition of nestling pied flycatchers in Wales. They began by devoting work during one breeding season to verifying that, indeed, growth bars denote 24 hours' growth of primary feathers. In 1996, they measured at 3-day intervals the growing second primary of the left wing, while, in 1997, they repeated the procedure with the third primary. When the nestlings were 15 days old, Kern and Cowie collected the primary feather they had been measuring. They counted the growth bars on as great a length of a feather as possible and divided that length of feather by that number of growth bars to determine daily feather growth according to the growth bar analysis. They then compared daily feather growth from the width of growth bars with the daily growth as calculated from measurements of the feather growing on the bird.

The results were surprising. On the primaries of 1996 and 1997, respectively, the width of growth bars was only 47% and 44% of the daily growth rate calculated from measurements on the bird. That is, there were about two growth bars per 24 hours, not one. Michael Kern was then on the biology faculty of The College of Wooster, Ohio, only a short drive from my university in Columbus. I was so intrigued when I was sent the manuscript of the article to review that I asked Kern if I could have a look at the feathers. I measured growth bars myself and, sure enough, they were about half as wide as the *in situ* measurements said they should be.

This is the third time we've encountered feathers with growth bar width different than expected from the 24-hour rule, and the first concerning nestlings. Langston and Rohwer (1996) found growth bars to be half as wide as predicted by the 24-hour rule. They knew that albatrosses have two major periods of activity during the non-breeding season, at dawn and dusk, and concluded that the two growth bars per day must relate to the two activity periods per day. Bob Mauck and I (1995) were not able to detect any pattern at all in the seemingly chaotic arrangement of light and dark segments on Leach's storm-petrel rectrices. We hypothesized that during the nestling-feeding period, when the induced feathers were being grown, the parent petrels had an irregular schedule of activity and rest periods depending on a complex interaction of the day/night cycle, the tidal cycle, and whether they were making a long-distance commute to and from the nesting burrow.

The present case with flycatcher nestlings seems to have no ready explanation. We might hypothesize that the nestlings have a bimodal physiological cycle per 24-hours, causing production of two growth bars. Also, similar to concerns with Murphy and King's (1991) white-crowned sparrow project, perhaps physically handling the nestlings every third day created an artifact of feather growth. Comparison of growth bar width of handled and non-handled nestlings would be useful. Finally, the anomalous result with the flycatchers may hint at some presently unknown biological rhythm in nestling birds. Comparative work with other species would be very valuable in assessing this possibility.

The other section of Kern and Cowie's paper on pied flycatchers concerned growth bar widths on two feathers little studied previously. One was the third tertial, one of the feathers between the secondary feathers of the wing and the body. The other was the sixth greater secondary covert, one of the feathers covering the upper side of the base of the secondary feathers on the wing. They suggested that since these two feathers are both molted during the winter (or prenuptial) molt, there should be a positive correlation between the widths of the growth bars they contain.

The analysis revealed that there was no relationship at all between growth bar widths of the two feathers. Kern and Cowie suggested that this lack of correlation spread doubt about the validity of growth bar widths as indices of nutritional condition. The difficulty with that suggestion is that this tertiary and this covert may actually be replaced during different parts of the winter molting period (E. V. Pravosudova, personal communication), so the nutritional circumstances of a bird may change between molting the first and second feather. As always, the best way to resolve an uncertainty like this is with a controlled test, in this case inducing both feathers at the same time in two groups, one food-deprived and the other a control.

9.1 *Summary*

This short chapter reviewing articles on feather growth in nestlings has brought forth two major points. From the swallow paper, we learned that feather growth appears to integrate nutritional condition over several days. We might imagine daily feather growth rate as a moving average through the time of feather development. Details of the period integrated and how the impact of any day's conditions might decay with time remain to be worked out. As a practical matter, we may not expect growth bar widths to be useful in detecting sharp, short changes in nutritional condition, although this idea, too, can be tested experimentally.

The nestling pied flycatcher result remains completely unexplained. It will be very useful to have records from other species, particularly box-nesting species where such records are quite easy to assemble. The potential artifact of handling needs to be evaluated.

10 *Prolonged brood-care*

10.1 *Adult nutritional condition*

Cooperative breeding in birds occurs when more birds than just two breeding adults participate in an attempt at reproduction (Stacey and Koenig 1990). In most cases, such cooperative groups consist of a breeding pair and a variable number of "helpers." Almost always, the helpers are the breeding pairs' offspring from former years that have remained in their natal territory rather than dispersing to find a mate and territory of their own. Although many studies have dealt with breeding season aspects of this social phenomenon, Jan Ekman has taken another tack by focusing on the non-breeding season (Ekman and Rosander 1992; Ekman *et al.* 1994). Ekman's position is that before we consider events during the breeding season, we should think about the mechanisms permitting or encouraging offspring to reside with their parents through the preceding non-breeding season, a time of limited resources. If such offspring were forced by their parents to leave their natal site, young-of-the-year birds would not be available to act as helpers the following spring.

The prolonged brood-care model (Ekman and Rosander 1992; Ekman *et al.* 1994) holds that parental control of the dispersal of offspring can be an important factor in determining the size and composition of winter groups of related animals. According to this model, if food resources are low, socially dominant parents wintering on their former breeding site do best by being competitive and reserving all resources for themselves. In a situation where food resources are more abundant, the model predicts that sharing resources with independent offspring would be favored by natural selection. Underlying such reasoning are the assumptions that delayed dispersal of offspring is linked to relaxed winter competition for food and that offspring have a higher probability of surviving their first winter if they stay in their parents' territory than if they disperse. Even with the existence of this bias toward their own young in the territorial pairs' behavior, the model assumes that there will be a cost to adults from sharing the limited resources in their territory with their young. That is, in order to retain their offspring through the winter, adult birds should have to sacrifice some resources, resulting in lower levels of nutrition for themselves. In contrast, territorial adults are expected to increase their level of aggressiveness toward non-related flock members and force them to leave. The model suggests that non-related young will be allowed to stay only if resources are sufficient to be shared with more than one additional flock member. Descriptive studies of Siberian jays (Ekman *et al.* 1994) and gray jays (Waite and Strickland 1997) have been consistent with the model in demonstrating that these two species of the crow family (Corvidae) favored their own retained offspring over immigrants in winter groups. Recently,

Elena Pravosudova and I employed ptilochronology in the first controlled manipulative test of the model.

In winter, tufted titmice form coherent conspecific groups of two to eight individuals (Grubb and Pravosudov 1994b). Such groups usually include two adult birds (a territorial pair) and, often, one or more of their offspring and/or unrelated first-year individuals from unknown natal sites (Pielou 1957; Brackbill 1970; Brawn and Samson 1983). Family groups extend into the breeding season, and helping at the nest has been reported (Pielou 1957).

In our study area within the agricultural landscape of central Ohio, tufted titmice wintering in forest fragments form small conspecific groups of two to five birds. Based on Ekman and Rosander's (1992) model, we predicted that if all young birds were removed from a winter group residing in a small woodland fragment, the remaining adults would not have to share the non-renewing food supply through the winter and therefore would be in better nutritional condition than territorial adults in an unmanipulated control group containing at least one retained offspring of the pair.

During the winters of 1995–96 and 1996–97, we captured 65 titmice (34 adults and 31 juveniles) from 17 different groups in 10 forest fragments. All woodlots were of approximately the same size, topography, and vegetation, and each was completely isolated from other woodlands by cultivated fields. In late November, each titmouse group consisted of an adult pair and one to three first-year birds.

Each woodlot was randomly assigned to the experimental or control group. During early winter, we mist-netted titmice at sunflower-seed feeders, but to minimize any effect of food supplementation on the nutritional condition of the adult titmice, we maintained feeders in woodlots for only the time required to attract and capture the birds. This was less than a week in every case. To determine genetic relationships, we took a blood sample from each bird in each social group.

While capturing adults, we banded and then removed all the young birds from the treatment woodlots and released them approximately 50 km away in the "bird feeder belt" of suburban Columbus. None of these young birds was seen again in the study area.

We used ptilochronology to compare the nutritional condition of adult titmice in the experimental and control woodlots, pulling the left and right outermost rectrices and then collecting the induced replacements 6 weeks later. To determine if the removal of young had an effect on the nutritional condition of adults, we measured growth bar width, feather mass, and total length of both original and induced feathers from all recaptured adult birds. Feathers were measured in a "blind" fashion, and for each dependent variable, we used the average of the values for the left and right feathers.

In preparation for statistical analysis, we took several precautions to avoid pseudoreplication, the improper inflation of sample sizes caused by assuming independence of non-independent samples. In the two cases where a woodlot contained two social groups of titmice, we randomly selected only one group for analysis. In cases where we collected induced feathers from both members of an adult pair, we randomly selected the feathers from either the male or the female for analysis. Five woodlots were used during both winters, but three of them contained new adult pairs during the second winter, pairs we considered independent of those in the first winter. For the two woodlots that had the same adult birds during both winters, we randomly selected only one year's data for analysis. Thus, the primary sampling unit was one member of one territorial pair per woodlot-year with all birds used only once.

We used analysis of covariance to determine the effect of removal of young on feather growth of adults. Treatment and sex were entered as factors, and to control for the effects of bird size, we used dimensions of original feathers as covariates. As I have reported earlier, we now know that original feathers of all birds cannot be considered to have been grown under equal nutritional conditions. Now recommended is using tarsus length to control for differences in bird structural size. However, a beauty of randomly assigning woodlots to treatment groups is that we can assume there is no bias. That is, we have no reason to think that the average adult in either the experimental or control group was better nourished or of higher individual quality to begin with. To determine relatedness among members of each social group, we used DNA fingerprinting.

Because the molecular analysis was performed only after the field manipulation, we could not be aware of the degree of relatedness between adult and first-year members of the various groups at the time we removed juveniles. Therefore, to test the model's main assumption, that parents incur a cost by permitting their young to remain with them throughout the winter, we initially used adults that at the beginning of the experiment had at least one retained offspring on their winter territory (kin-group adults). We first compared nutritional condition of control adults that shared their territory with at least one retained offspring with nutritional condition of experimental adults from whose territories such offspring (along with unrelated young, if present) had been removed. A total of eight recaptured birds (two from experimental and six from control groups) were available for this analysis. The differences between experimental and control groups were insignificant for growth bar width, feather mass, and feather length, although all three measures of feather growth were somewhat greater in control birds.

To see if this trend persisted with larger sample sizes, we added records from experimental adults that initially had lacked retained offspring in their social group (non-kin adults). For a number of reasons, combining adults in this way seemed justified. First, because of the removal, neither kin-group nor non-kin experimental adults were required to share resources with any first-year birds. Second, the original tail feather measurements of kin-group and non-kin adults did not differ significantly for any of the three measures, growth bar width, feather mass, or feather length, indicating that the two groups of adults were of similar quality. Third, we could not detect any difference in induced feather measurements between the two kin-group experimental adults and five non-kin experimental adults.

The comparison of experimental and control groups performed after lumping records for the two kinds of experimental adults reinforced the trend existing in the earlier comparison. Contrary to the prediction from the model, average growth bar width of induced feathers in experimental-group adults was significantly narrower, not wider, than average growth bar width of controls (Figure 10.1a). Induced feathers from experimental birds were also significantly lighter than those from controls (Figure 10.1b), but the difference between treatment and control feathers was not significant for total length.

In kin-group adults, we found no difference between the growth bar width, mass, or total length of induced feathers of four males and four females standardized for average values of original feathers. After increasing the sample size by adding values from non-kin experimental birds, we had available records from eight females and five males. Average induced growth bar width standardized for average growth bar width of original feathers was significantly greater in males, as was standardized average total length, but not standardized average feather mass. There was no significant two-factor interaction between treatment and sex for any of the three dependent variables.

The prolonged brood-care model assumes that sharing resources with kin young has a cost for adult birds. Based on this assumption, we predicted that adults from manipulated groups (young removed) would do better nutritionally than adults sharing food with their offspring. The results of our

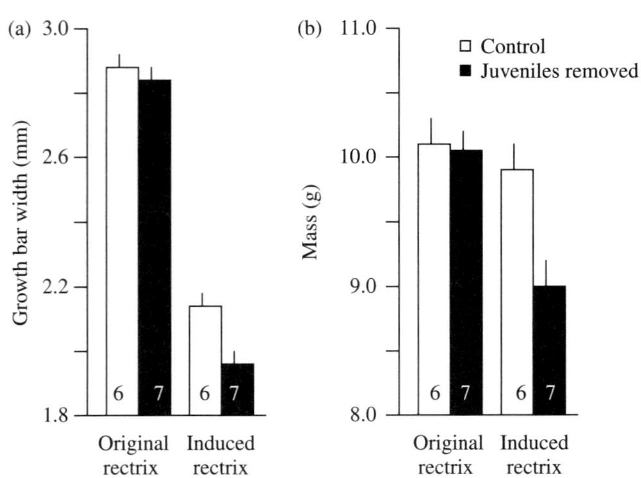

Figure 10.1 Mean ± SE growth bar width (a) and mass (b) of original and induced outermost pairs of tail feathers of adult tufted titmice wintering in the presence or absence of first-year titmice including one of their own offspring. Sample sizes are shown in the bars. From Pravosudova and Grubb (2000).

experiment suggest that this was not the case; instead, such adults seemed to have done worse. There might be several reasons for this disparity between prediction and result. First, even under a limited food supply, a larger group size may confer a fitness advantage. For an adult, the benefits of improved foraging efficiency due to lowered vigilance and better predator detection may outweigh the costs of sharing resources. Second, a deciduous forest habitat such as our study area may provide a relatively high winter supply of food. In deciduous habitats, territorial birds might not have been selected to be as despotic toward subordinate flock members as would be the case in species living in more harsh northern coniferous woodlands (Ekman *et al.* 1994; Waite and Strickland 1997). The adaptiveness of maintaining a larger flock size may outweigh that of monopolizing scarce resources. Thus, resident adult tufted titmice may tolerate one or more juveniles in their winter territory without significantly sacrificing their food supply.

Although, because of the model's assumptions, control groups in which all young were unrelated to adults had to be excluded from the test of the prediction, we did use adults from treatment groups that had contained only non-kin young before manipulation to increase our sample size. We showed that kin-group and non-kin experimental adults did not differ significantly in nutritional condition during the experiment. This result is not likely due to small sample sizes because, for the same group of treatment adults, the differences between males and females for all three feather parameters were highly significant, with all values being higher in males.

Regardless of the reasons that some groups lacked kin young in early winter (e.g. adults' reproductive attempt had failed or all resident young had dispersed), analysis of original feathers indicated that the nutritional condition of adult titmice in all the woodlots was about the same at the start of the manipulation. Whether adults spending the winter with retained offspring differ in nutritional condition from adults spending the winter in non-kin groups is beyond the scope of our original objective, but no induced feather growth parameter was significantly different between the seven kin-group and four non-kin control adults.

Male titmice grew their induced feathers faster and to a greater total length than did females, a result holding even after standardization to values of original feathers. In earlier work, David Cimprich and I (Grubb and Cimprich 1990) found a similar sex difference in feather regeneration in this species. As adult female tufted titmice are socially subordinate to adult males in winter flocks (Grubb and Pravosudov 1994b), it is not surprising that females in our sample were in comparatively poorer nutritional condition. However, since no significant interaction was shown between treatment and sex, there is no evidence that the manipulation affected sexes differently.

In conclusion, our results did not support the key assumption of the prolonged brood-care model. In contrast, our experiment demonstrated that the nutritional condition of territorial adults improved, rather than suffered, in the presence of related conspecific first-year members of a winter flock.

Therefore, in some systems the nutritional benefits for adults of associating with young in winter may outweigh the costs of sharing food.

10.2 *Nutritional condition of retained offspring*

Elena Pravosudova and I also used ptilochronology as part of a project testing predictions of the prolonged brood-care model that focused on the offspring. According to Ekman and Rosander's (1992) model, during the winter months when food is limited, dominant territorial adults tend to monopolize resources, but they should be more willing to share such resources with their own offspring than with unrelated first-year immigrants from foreign territories. In winter social groups, adult Siberian jays were not aggressive toward their own offspring, but prevented immigrant young from sharing food (Ekman *et al.* 1994). Additionally, retained-offspring immature jays obtained greater food and spent more time at an artificial food source in their parents' territory than did immigrant first-year birds (Sklepkovych 1997). Overall, first-year survival in this jay species seemed to be higher in retained offspring than in immigrant young (Ekman *et al.* 2000).

Using tufted titmice, Pravosudova and I tested two predictions from the prolonged brood-care model: (1) first-year birds related to the dominant adult pair should be in better nutritional condition than unrelated immigrant young, and (2) territorial adults should be more aggressive toward unrelated, immigrant young than toward their own offspring.

During early winter, we mist-netted titmice at sunflower-seed feeders. Group size was stable throughout the winter and ranged from three to seven birds. To facilitate molecular genetic analysis of relatedness among the individuals in each flock, we took a blood sample before releasing each bird at the capture site.

To test the prediction that immature birds related to a territorial adult should be in better nutritional condition than unrelated immature birds, we plucked the two outermost rectrices (L6 and R6) from all first-year birds. Induced rectrices were collected from the same birds at the end of the winter at least 6 weeks later. To minimize the effect of food supplementation on the nutritional condition of the birds, we maintained feeders in woodlots for only the time required to attract and capture the birds, less than a week in every case.

To determine if kinship had an effect on the nutritional condition of young, in a blind fashion, we measured growth bar width and total length of both original and induced feathers from the 13 young birds we recaptured. For analysis, we used the average values of left and right feathers and employed tarsus length to control for differences in the structural size of birds.

In the comparison of feather growth, we found no significant difference between kin and non-kin young for either induced growth bar width or induced total feather length. However, average bar width and total length of original feathers were both significantly greater on kin young (Figure 10.2).

To test the second prediction, that adults should be more aggressive to immigrants than to retained young, we caught and banded the birds in each

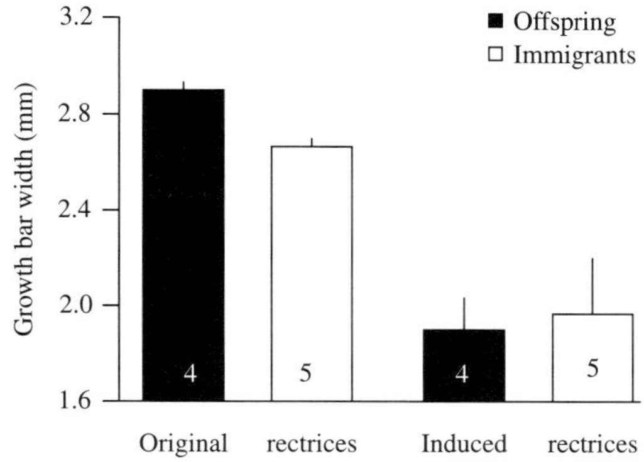

Figure 10.2 Mean ± SE growth bar width of original and induced tail feathers of offspring and immigrant immature tufted titmice. Sample sizes are shown in the bars. From Pravosudova *et al.* (2001).

flock, then replaced the sunflower-seed feeder with a new bait, a fine-mesh nylon bag filled with shelled whole-kernel Spanish peanuts. The peanuts could not be easily extracted through the mesh, so titmice had to stay on a bag and peck for several minutes, a necessity which provoked interactions among members of a group. We recorded interactions at the feeding site using the four-category system of Ekman *et al.* (1994). Because genetic analysis of relatedness was performed after the field season, our observations were not biased by knowledge of the genetic similarity among the members of each group.

We had records from four adult/offspring pairs of titmice and from seven adult/non-offspring-young pairs. Adults were significantly more aggressive toward immigrants than toward their own young. Intriguingly, there was a significant negative correlation between the coefficient of molecular band-sharing, a statistic indicating how closely related two birds are genetically, and the aggression score for adults and immature birds interacting at the bait (Figure 10.3). We know from other work that in titmice, first-order relatives, in this case a parent and an offspring, have molecular band-sharing values of at least 0.46. Figure 10.3 suggests that adults seemed to be discriminating among young to which they were less related than to their own offspring. To what extent an adult titmouse can distinguish between, say, a niece or nephew and less-related non-offspring young is unknown.

Contrary to our first prediction, growth rates of induced feathers suggested that kinship was unlikely to have an effect on the nutritional condition of first-year tufted titmice. This finding is at odds with behavioral observations on Siberian jays (Sklepkovych 1997), which showed better access to food by retained offspring. As mentioned in conjunction with our study of the nutritional condition of adult titmice, it may be that the lower latitude deciduous woodlands in which we worked are less stressful nutritionally than are the higher latitude coniferous woodlands of the Siberian jay study. A second factor, and one we tried to minimize, is that our study was potentially

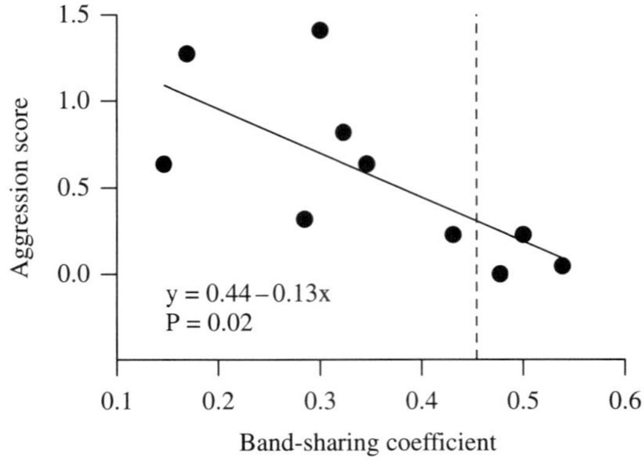

Figure 10.3 Relationship between band-sharing coefficient and aggression score in two-bird comparisons between territorial adult and immature tufted titmice in winter groups. The dashed line indicates the threshold band-sharing coefficient ($x = 0.46$) for first-order relatives such as parents and offspring. Adults were progressively more aggressive toward progressively more distantly related immature birds. From Pravosudova *et al.* (2001).

confounded by an artifact, namely access to supplemental food by non-kin first-year titmice. In order to catch the birds for color-banding, we pre-baited them to a sunflower-seed feeder for several days. During that period, we assume the titmice in a flock would have cached seeds in the surrounding woodland. We do not know how many seeds each bird cached during those few days, but if non-kin young had access to self-cached seeds during the feather re-growth period, such supplementary food could have increased their nutritional condition sufficiently to overcome any adverse effect of adult aggression around the peanuts. Future work with this and other species must employ methods of catching birds for banding and feather plucking that do not rely on potentially confounding supplementary food.

Analysis of original titmouse feathers suggested that offspring were in better nutritional condition during the nestling/fledging period than were immigrant young. The significant differences in growth bar width and total feather length were present in the same birds that did not seem to have differences in induced feather variables during the winter. The differences in original feather variables were not due to differences in overall size as the analysis controlled for size variation by standardizing against tarsus length.

In species with pronounced delayed dispersal, not only do juvenile males tend to stay on the natal territory beyond fledging more often than juvenile females (Tarboton 1981; Stacey and Ligon 1987), but also in some species, dominant juveniles force subordinate siblings from the natal territory (Strickland 1991). Our records support the idea that socially dominant immature birds in winter may have been in better nutritional condition at the end of summer and may have been the ones that delayed dispersal and stayed in the winter group with their parents. Thus, it is possible that in some systems, kin young are better off during the winter not due to prolonged parental care, but rather due to the fact that they were higher quality juveniles to begin with.

Overall, our results partially supported the prolonged brood-care hypothesis. The lack of complete support of our predictions may also have occurred because factors other than kinship partially determine nutritional condition and social interactions among members of winter groups. Unlike many corvid and woodpecker species that typically delay dispersal, in the winter, tufted titmice occur in mixed-species foraging flocks where they are socially dominant to other species such as Carolina and black-capped chickadees (Dolby and Grubb 1998; Pravosudov and Grubb 1999). Therefore, even the most subordinate bird in a conspecific group of titmice is still dominant over birds of other species foraging in the same flock. Overall, it is possible that parental tolerance in this titmouse during winter has more to do with long-term inclusive fitness benefits for territorial adults and with long-term direct and indirect fitness benefits for offspring than with short-term nutritional costs.

10.3 *Summary*

This chapter has focused on the only two studies employing ptilochronology to test predictions stemming from the hypothesis of long-term brood-care. Both for adult and for juvenile tufted titmice, nutritional condition, as indexed by rate of induced feather growth, did not support the hypothesis. Pravosudova and I advanced a number of possible reasons why our results were not positive. Additionally, it seems possible that the nutritional advantage gained by offspring juveniles over non-offspring juveniles in a winter group may have been too small to be detected by differences in induced feather growth. Such an idea, however, does not explain why the relative feather growth rates of adults with and without juveniles in their territory should be statistically significantly opposite that predicted by the theory. It is too early to draw firm conclusions about the generality of the prolonged brood-care theory based on experimental work from just one laboratory, ours. We must wait for projects of a similar nature, but with other systems. In particular, it would be quite useful to have these predictions about nutritional condition tested with ptilochronology on Siberian jays, the species that inspired the hypothesis in the first place.

11 *Taking stock and looking ahead*

11.1 *Conceptual issues*

In the previous 10 chapters, we have considered the current state of basic and applied ptilochronology. Chapters 1–4 explored attributes of the method, how it relates to other measures of nutritional condition, and reviewed a number of criticisms of the technique. The six subsequent chapters dealt with applications of ptilochronology to a number of conceptual areas of behavior, ecology, and conservation biology.

As each of the preceding 10 chapters ended with a summary, to simply add another here would be redundant. Instead, this chapter brings forward conceptual issues that span various topics, and highlights uncertainties needing to be investigated in the future.

11.2 *What causes growth bars?*

It is obvious that some growth bars are characterized by regular variation in extent of pigmentation. Rectrices of house wrens and Styan's grasshopper warblers (Figure 1.2c) are examples of feathers showing regular variation in extent of deposition of melanin. We do not know, however, whether the more heavily pigmented band of each growth bar is laid down during the day or night.

Pigmentation does not explain the presence of perceptible growth bars on monochromatic feathers. That the regular progression of alternating lighter and darker bands on such feathers is most noticeable when light is directed obliquely onto the feather and reflected into the observer's eye suggests that growth bars are caused by a subtle undulation of the feather surface. What could cause such an undulation? One suggestion yet to receive investigation is that birds have circadian cycles of metabolic rate, with rate higher in the daytime, and that such variation causes the undulation underlying growth bars (Riddle 1908). The logic is as follows. The circadian variation in metabolic rate causes slight variation in turgor pressure of the blood within the keratin sheath surrounding the growing feather. From knowledge that metabolic rates of some birds are known to be lower at night during the roosting period (e.g. Reinertsen and Haftorn 1986), we might assume that the turgor pressure will be lower at night than during the day. The lower turgor pressure at night will exert less pressure on the follicle, causing the diameter of the pulp cavity to contract slightly then. Because of the way a feather vane grows while folded within the pulp cavity against the inside of the sheath, the circadian variation in pulp cavity diameter will lead to circadian variation in the orientation of the vane relative to the shaft or rachis of the feather. For example, parts of the vane of a tail feather grown at night should be slightly lowered dorsally compared with parts of the feather grown during the day.

Stretch-receptive cuffs could be applied around the pulp-containing sheath of a growing feather and connected to a data-logger mounted, for example, on the bird's tail. If such were done, one might predict a 24-hour rhythm in sheath diameter as measured by the stretch receptor.

Growth bars are harder to detect on large feathers, and become essentially undetectable on very large feathers such as the primaries of Canada geese or turkey vultures. Growth bars on smaller feathers, for example secondary coverts, of the same large birds are readily noticeable. There could be two reasons for this difference, neither of which has been examined. First, the difference between day and night metabolic rate and, therefore, the difference in turgor pressure within a feather sheath may be less in larger birds. Second, the sheath around large feathers may be thicker and less flexible, making it less responsive to the circadian changes in turgor pressure within the pulp cavity. Hopefully, these and other possible proximate causes of growth bars will come under investigation in the near future.

11.3 *On 24-hours' worth of feather growth per growth bar*

The early measurements on house finches (Wood 1950) and later radio-assays of willow tits (Brodin 1993) both suggest that ascribing one growth bar to 24-hours of feather growth is a safe procedure. And yet, we have encountered three troublesome exceptions. Bob Mauck and I (Mauck and Grubb 1995) were unable to make any sense out of the chaotic pattern of darker and lighter "bands" on Leach's storm-petrel rectrices, and Langston and Rohwer (1996) assumed they were viewing two growth bars per 24-hour period on albatrosses. In both cases, the atypical barring patterns were ascribed to atypical variation in activity and rest periods of these pelagic seabirds. Kern and Cowie's (2002) finding of two growth bars per 24-hour period on the primary wing feathers of pied flycatcher nestlings is both puzzling and intriguing. Might the flycatcher nestlings have two periods of activity and two of inactivity per 24-hours, marked by two levels of turgor pressure in their feather-pulp cavities? Could nestling birds have some previously unknown strategy for conserving maintenance energy? Metabolic study and growth bar examination of nestlings of other species should draw us closer to answers.

11.4 *Linking growth bar width and nutritional condition*

Fundamental to ptilochronology is the assumption that a reduction in net intake of energy causes a reduction on feather growth rate and, therefore, growth bar width. In the years since Murphy and King (1991) questioned this assumption, buttressing evidence has accumulated. Controlled laboratory studies (Grubb 1991; Jenkins *et al.* 2001) have shown that restricted diets cause narrower growth bars. Field experiments have revealed that birds permitted to retrieve supplemental food that they had cached previously grew induced feathers with wider growth bars than did controls (Waite 1990; Nilsson *et al.* 1993). Strong and Sherry (2000) contributed correlational evidence, a positive relationship between amount of ant prey on the territory of wintering ovenbirds and the growth rate of the birds' induced rectrices.

Much less clear is the issue of dose-dependency, whether a predictable quantitative relationship exists between nutritional condition and feather growth rate. We know that chickadee diets calibrated at 90% of *ad libitum* did not slow feather growth significantly compared to 100% controls, but 80% diets did (Grubb 1991). We also know that under field conditions, some birds delayed or forewent regenerating an induced feather (Waite 1990; White *et al.* 1991; Montiero and Furness 1996). At this point, how finely nutritional condition can be calibrated from feather growth rate remains an open question.

11.5 *Potentially confounding factors*

How tight is the cause-and-effect relationship between nutritional condition and feather growth rate? Stated another way, how vigilant must we be for the possibility that some factor other than nutritional condition could be affecting feather growth rate and confounding our interpretation of results? Jenkins *et al.* (2001) thought they had detected a generality, that feather growth rate could be influenced by density-dependent ecological factors, but not by density-independent factors. For example, they considered dominance position within the flocks of sparrows they studied to be density dependent, which it is, and showed there was a relationship between dominance and feather growth rate. Presumably, Jenkins *et al.* (2001) had in mind that density-dependent factors and feather growth rate would be negatively correlated, an increase in density or density-related factor causing reduced feather growth. At first consideration, it seems that Brown *et al.* (2002) furnish contradictory evidence, with growth bar width on wintering hermit thrushes highest in the habitat with the highest thrush density. However, fruits were also most abundant in that habitat, so one could argue that density of thrushes per unit of fruit was actually lowest in the habitat with the highest density of thrushes.

The idea that density-independent factors do not affect feather growth rate quickly runs into counter-examples. Snow cover, a density-independent factor, reduced induced feather growth in tufted titmice (Doherty and Grubb 2003). Similarly, day length in American tree sparrows (White and Kennedy 1992) and administered doses of testosterone in starlings (De Ridder *et al.* 2002), both density-independent factors, were shown in controlled studies to influence feather growth rate.

In a pair of experiments, two density-independent factors, the previous history of feather generation by a follicle (Grubb and Pravosudov 1994*a*) and the wind-chill regime under which birds on *ad libitum* food regenerated a feather (Zuberbier and Grubb 1992) were shown not to affect rate of feather growth. It is important to realize that these results were contingent on the particular circumstances of the experiments. Growing feathers more continuously from a given follicle or under even colder conditions, for example, could very well have an effect on growth rate and, therefore, growth bar width.

It has been universally true that growth bar width and total length of induced feathers are narrower and shorter, respectively, than those of original feathers grown during the normal fall molt, and this has been the case even when the induced feathers were grown in the presence of *ad libitum* food (e.g. Grubb and Pravosudov 1994*a*). This difference cannot be due to density-dependent factors alone, and may result, in part, from different cellular processes involved with normal growth versus tissue repair (Talloen *et al.* 2005). Clearly, manipulative experiments should employ experimental and control replicates contemporaneously.

11.6 *Original versus induced feathers*

Many of the projects discussed in previous chapters used growth bar width on original feathers as a measure of nutritional condition. The extent of the danger of misinterpretation of results from original feathers depends on the particular circumstance of the project. At the "safe end" of the continuum is a project such as our study of nutritional condition in Florida scrub-jay fledglings (Grubb *et al.* 1998) where we were certain that each young bird had grown its original feather in the location where that feather was collected. Less certain are results from Yosef's and my assessment of nutritional condition in wintering loggerhead shrikes (Grubb and Yosef 1994). We made the assumption that the territorial adult shrikes we sampled had grown the original feather at the capture site during the previous molt, but we could not be certain this was true. The projects of Carbonell and Telleria (1999) and Stratford and Stouffer (2001) are even more problematical as some birds were known to have grown the focal original feather somewhere other than the capture location. Nevertheless, use of original feathers, despite uncertainty about their provenance, can be quite enlightening, as demonstrated by Carlson's comparison of historical and contemporary feather growth rates in white-backed woodpeckers (Carlson 1998).

11.7 *Controlling for structural body size*

All else being equal, a bigger bird will grow a longer feather with wider growth bars. Coping statistically with this potentially confounding relationship continues to be difficult. For quite a few years, workers relied on the assumption that during the presumably halcyon days of late summer and early autumn, feather growth rates were limited only by the ability of follicles to assemble and incorporate materials into the growing feather. Alternatively, a fallback assumption was that if there were some environmental limitation to the rate of feather growth during the normal molting period, all study birds would be limited to the same extent. Thus, we could use the ratio of induced to original feather growth rate to control for body size or, more recently, employ the rate of original feather growth as a covariate in multivariate analyses.

Several projects in the chapter on individual quality showed variation among birds in growth rate of original feathers (e.g. Hill and Montgomerie 1994; Keyser and Hill 1999; Takaki *et al.* 2001) and clearly put the kibosh on using original feather growth in descriptive studies to standardize for body size. At the moment, there seem to be two possible solutions to the problem. The first is to employ some measure of skeletal size, such as length of tarsometatarsus, as a variable in multivariate models. As I have mentioned several times already, the other solution to the standardization problem is to randomly assign birds to treatment groups. Assuming that sample sizes are sufficiently large, variation in structural size will then contribute to within-group variance in feather growth rate, but have no effect on between-group variance.

11.8 *Feather growth lag times*

Particularly important for descriptive studies is knowledge of how long a time period affects feather growth. Brodin's (1993) radio-ptilochronology project with captive willow tits employed the assumption that the radioactive sulfur in ingested sunflower seeds was incorporated into the growing rectrix very quickly, on the order of hours. Certainly, the onset of radioactivity, as depicted by radiographs, seemed to occur over a very brief period, possibly on the order of minutes. There was no suggestion of a gradual incorporation over the course of days. By contrast, though McCarty and Winkler (1999) measured feather growth on the bird rather than from growth bar widths, their statistical analysis indicated that conditions as much as several days before could have influenced rate of feather growth at any given time. Adult willow tits versus nestling tree swallows; measured growth bar width versus measured growth *in situ*. Different methods and different results. Resolution might be to see how tightly on an induced feather wide growth bars are concentrated in response to a brief pulse of super abundant food given to birds maintained on less that *ad libitum* food before and after the food pulse. Something like such a follow-up to my chickadee experiment (Grubb 1991) should provide an answer.

11.9 *Relations among growth bar width, feather length, and feather mass*

All else being equal, one would think that growth bar width, feather total length, and feather mass would all respond similarly to nutritional condition. In good times, all would be greater and in lean times, smaller. Such has not been the case. In assessing the impact of various environmental factors on the three dependent variables, some projects have found significant effects on growth bar width only (Dolby and Grubb 1998), on total length only (Jenkins *et al.* 2001), or on feather mass only (Carbonell and Telleria 1999). Some have found significant relationships with growth bar width and feather mass, but not feather length (White *et al.* 1991; Pravosudova and Grubb 1999), and with feather length and mass, but not with growth bar width (White and Kennedy 1992). Some of this variation may be due to measurement accuracy. For example, feather length is a function of growth bar width times the number of days the feather is grown, so it might seem reasonable to conclude that differences in feather length and mass between, say, treatment groups would be easier to detect than differences between growth bar width. Alternatively, different mechanisms may control daily growth and total period of growth. White and Kennedy (1992) concluded from their study of photostimulated American tree sparrows that the number of days a feather grew was controlled by hormones keyed to day length, while the rate of growth per day was controlled by nutritional condition. The cases where daily growth, but not feather length, differed between treatment groups suggest that some factor or factors operate at the level of the follicle to shut off further growth. Does a follicle somehow monitor the cumulative amount of material and/or energy it is devoting to growing a feather and shut down production when some target accumulation is reached?

11.10 *Ptilochronology and fitness*

Of paramount interest to evolutionary ecologists is whether feather growth rate can serve as a sensitive indicator of forces that affect fitness. In 1991, Mary Murphy and James King correctly pointed out that this central assumption behind the utility of ptilochronology was unsupported by evidence. We now have a handful of projects linking feather growth rate to widely recognized fitness components. Those habitat patches promoting higher chickadee annual survivorship (Doherty and Grubb 2002) also produced birds with wider feather growth bars (Doherty and Grubb 2003). Hermit thrushes wintering in the habitat with the highest daily feather growth rates were most likely to return the next winter (Stratford and Stouffer 2001). In a manipulative project, habitats containing two nuclear species of mixed-species flocks produced white-breasted nuthatches with both wider growth bars on induced feathers and a strong though not significant tendency to better survive the winter (Dolby and Grubb 1998).

Turning to the reproductive component of fitness, wider feather growth bars indicated that brighter male house finches were in better nutritional condition. The brighter coloration appeared to be an honest signal of nutritional condition used in mate selection by females (Hill 1991; Hill and Montgomerie 1994). Male blue grosbeaks with wider feather growth bars indicating better nutritional condition were also brighter blue. Females selecting bright blue males as mates would be selecting males more likely to bring food to the female and nestlings, so growth bar width was positively associated with feeding rate of offspring (Keyser and Hill 2000). In the highest-quality evidence to date, Takaki *et al.* (2001) showed significant correlations between growth bar width and both survivorship and reproductive success in a grasshopper warbler, the only study demonstrating linkages between feather growth rate and both major components of fitness.

We now have additional support for an early theme, the idea that growth rate of induced feathers is a better indicator of nutritional condition than are other more traditional measures such as body mass or fat score. While subordinate "floater" Norwegian willow tits had significantly reduced rates of induced feather growth compared with the most subordinate member of flocks, fat supply was uninformative; there was no difference between the two categories of birds. Contrary to the conclusion of Carrascal *et al.* (1998), I suggest that it was the growth bar widths that were informative about the nutritional condition of tits in different habitats (narrower growth bars in the colder, less food-rich site), not pectoral muscle (which did not differ between sites). Ovenbirds wintering in Jamaica had the narrowest induced growth bars in the driest of three study habitat types, the one known to have the lowest density of the birds' ant prey. By contrast, birds in the dry, ant-poor habitat type carried the most fat. Greatest fat supplies in the poorest-quality habitat is just what might be expected if birds with less predictable resources carry a larger insurance policy against starvation (Rogers 1987; Rogers and Smith 1993).

11.11 *Ptilochronology and conservation biology*

Chapter 4 brought forward a number of examples where rate of feather growth had been employed in the context of conservation biology. The Belgian study of nutritional effects of lead, in particular, points to the potential of feather growth rates as a bioassay. The same project cast doubt on the potential of fluctuating asymmetry as an index of stress, nutritional or otherwise.

Results from ptilochronology completely reversed conventional wisdom about the relative quality of two pine habitats for the endangered red-cockaded woodpecker. This example, in particular, indicates a bright future for feather growth rates as indicators of habitat quality useful to conservation biologists.

In the realm of conservation biology as well as in the other conceptual areas discussed in this book, we await more results from applications beyond North America and Eurasia. The only examples so far from other parts of the world are Strong and Sherry's (2000) study of Jamaican ovenbirds, Stratford and Stouffer's (2001) focus on forest fragmentation in the Amazon, Spanhove and Lens's (unpublished) study of white-starred robins in Kenyan woodland fragments, Velando's (2002) Peruvian booby project, and Doucet and Montgomerie's (2003) work with Australian bowerbirds.

11.12 *Correlation versus causation*

My final point is to focus once again on the importance of manipulated projects with random assignment of replicates to treatment groups. Whenever feasible, investigators should invest time and effort in manipulations, sacrificing sample sizes if need be. Such a policy will become considerably more attractive if journal editors recognize that the much higher quality of data from manipulative experiments compensates considerably for the necessarily reduced number of replicates obtained. Having said that, I await the new insights into evolution, ecology, and conservation biology to be produced by further explorations with ptilochronology.

Appendix 1. Ptilochronology measurements and methods

The general technique for obtaining ptilochronology feather measurements starts with capturing a bird and plucking a feather, typically the outermost right tail feather, or rectrix. Removing a feather causes no visible trauma to a bird, likely because a fully grown feather is held in place in the follicle solely by non-living connective tissue. After being so plucked, the bird is released, recaptured after the 5–6-week growth period of the induced feather, and the induced feather pulled. Feathers are usually stored individually in envelopes. (A fully grown feather has no pulp remaining in the calumus (or shaft) and the sheath that surrounded the feather while it grew has been fully preened away by the bird.)

My original method for measuring the width of growth bars involves first taping a rectrix by the base (calamus) of its shaft to an index card (Grubb 1989). I then push a size 0 (i.e. very thin) insect-mounting pin perpendicularly through the feather and underlying card at the distal edge of each dark growth band, the edge that is closer to the feather's tip. I also pierce the underlying card at the proximal (base) and distal (tip) ends of the feather.

Because feathers often differ in the number of growth bars that can be seen clearly, I developed a standard technique to determine the width of growth bars. I use a caliper to measure the total length of the feather to the nearest 0.01 mm. Then, I calculate and mark on the card the point two-thirds of the feather's length from the proximal end. My normal procedure for most songbirds is to determine the average width of 10 growth bars. These include the growth bar on which the "two-thirds" point falls plus four proximal and five distal bars. I divide the total length of this 10-bar segment by 10 to obtain an average value of daily growth (Figure A.1). The basic assumption is that the greater the average growth bar width, the better the bird's nutritional condition while growing the feather.

Geoff Hill and his students (Shawkey *et al.* 2003) employed a gel documentation system to count and measure growth bars more efficiently. The instrument, designed to detect and measure bands of molecules on gels, takes digital photos of growth-bar-bearing feathers in a low-light chamber. Such photos are then enhanced and measured with a computer. While less time-consuming than the original system, this new approach requires a $15,000 gel reader. Thus, it is important to note that Shawkey *et al.* (2003) found the gel reader to be no more accurate than the original system that requires only an index card and a pin.

Using WinDENDRO software (http://www.regent.qc.ca/products/dendro/DENDRO.html), Kenneth Otter and Sarah Atherton (personal communication)

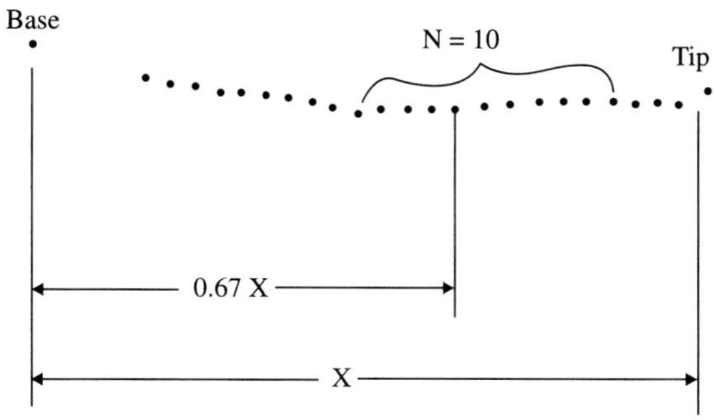

Figure A.1 Method for determining the width of a feather growth bar by measuring pinpricks punched through the feather at the edge of growth bars, and then through an underlying card. The 10 growth bars centered on the point two-thirds of the distance toward the tip of the feather are measured in the aggregate and then divided by 10 to determine average growth bar width.

are currently attempting to apply to feather growth bars technology developed for recognizing and measuring tree rings.

As an index of nutritional condition within a species, induced feather growth must be comparable across birds of different size. To obtain a standardized index, my original method was to determine the daily growth of the original feather (the first one plucked). I assumed that all such original feathers were grown during the previous molting season. I then divided the daily growth rate (mm/day) of the induced rectrix by the daily growth rate of the original rectrix to obtain the proportion of the growth rate during the normal molting period represented by growth rate of the induced feather. I concluded that, regardless of the size of the bird, this proportion of induced to original feather growth was positively correlated with nutritional condition. In later work, instead of using the proportion just described, I used the average width of growth bars on original feathers as a covariate in multivariate analyses. Most recently, other workers and I have taken to employing the length of a bird's tarsometatarsus, the long bone of the leg, to adjust for structural size of the bird.

Appendix 2. Scientific names of bird species mentioned

Common name	Scientific name
Albatross, black-footed	*Diomedea nigripes*
Albatross, Laysan	*Diomedea immutabilis*
Blackbird, red-winged	*Agelaius phoeniceus*
Blackcap	*Sylvia atricapilla*
Booby, blue-footed	*Sula nebouxii*
Cardinal, northern	*Cardinalis cardinalis*
Chickadee, black-capped	*Poecile atricapillus*
Chickadee, Carolina	*Poecile carolinensis*
Creeper, brown	*Certhia americana*
Finch, house	*Carpodacus mexicanus*
Goose, Canada	*Branta canadensis*
Grassquit, blue-black	*Volatinia jacarina*
Grebe, eared	*Podiceps nigricollis*
Grosbeak, blue	*Guiraca caerulea*
Jay, gray	*Perisoreus canadensis*
Jay, Siberian	*Perisoreus infaustus*
Junco, dark-eyed	*Junco hyemalis*
Manakin, white-crowned	*Pipra pipra*
Nutcracker, Clark's	*Nucifraga columbiana*
Nuthatch, European	*Sitta europaea*
Nuthatch, white-breasted	*Sitta carolinensis*
Peacock	*Pavo ocellatus*
Ovenbird	*Seiurus aurocapillus*
Robin, white-starred	*Pogonocichla stellate*
Scrub-jay, Florida	*Aphelocoma coerulescens*
Shrike, loggerhead	*Lanius ludovicianus*
Shrike, northern	*Lanius excubitor*
Sparrow, American tree	*Spizella arborea*
Sparrow, white-crowned	*Zonotrichia leucophrys*
Sparrow, white-throated	*Zonotrichia albicollis*
Starling, European	*Sturnus vulgaris*
Storm-petrel, Leach's	*Oceanodroma leucorhoa*
Tern, common	*Sterna hirundo*
Thrush, hermit	*Catharus guttatus*
Thrush, pallid	*Turdus pallidus*
Tit, blue	*Parus caeruleus*
Tit, coal	*Parus ater*

Continued

Common name	Scientific name
Tit, great	*Parus major*
Tit, marsh	*Parus palustris*
Tit, willow	*Parus montanus*
Titmouse, tufted	*Baeolophus bicolor*
Tropicbird, red-tailed	*Phaethon rubricauda*
Vulture, turkey	*Cathartes aura*
Warbler, graceful	*Prinia gracilis*
Warbler, grasshopper, Styan's	*Locustella pleskei*
Warbler, hooded	*Wilsonia citrina*
Woodcreeper, wedge-billed	*Glyphorynchus spirurus*
Woodpecker, downy	*Picoides pubescens*
Woodpecker, hairy	*Picoides villosus*
Woodpecker, red-bellied	*Melanerpes carolinus*
Woodpecker, white-backed	*Dendrocopos leucotos*
Wren, house	*Troglodytes aedon*

References

Anderson, D. J. and R. Ricklefs. 1992. Brood size and food provisioning in masked and blue-footed boobies (*Sula* spp.). *Ecology* 73:1363–1374.

Andersson, S. 1994. Costs of sexual advertising in the lekking Jackson's Widowbird. *Condor* 96:1–10.

Barnard, C. J. and D. B. A. Thompson. 1985. *Gull and povers: the ecology and behaviour of mixed-species feeding groups*. Croom Helm, London.

Bierregaard, R. O., Jr. and T. E. Lovejoy. 1989. Effects of forest fragmentation on understory bird communities. *Acta Amazonica* 19:215–241.

Birkhead, T. R. and A. P. Møller. 1998. *Sperm competition and sexual selection*. Academic Press, San Diego, CA.

Brackbill, K. 1970. Tufted titmouse breeding behavior. *Auk* 87:522–536.

Brawn, J. D. and F. B. Samson. 1983. Winter behavior of tufted titmice. *Wilson Bulletin* 95:222–232.

Breininger, D. R., V. L. Larson, B. W. Duncan, R. B. Smith, D. M. Oddy, and M. F. Goodchild. 1995. Landscape patterns of Florida scrub-jay habitat use and demographic success. *Conservation Biology* 9:1442–1453.

Brodin, A. 1993. Radio-ptilochronology—tracing radioactively labeled food in feathers. *Ornis Scandinavica* 24:167–173.

Brodin, A. and J. Ekman. 1994. Benefits of food hoarding. *Nature* 372:510.

Brown, D. R., C. M. Strong, and P. C. Stouffer. 2002. Demographic effects of habitat selection by hermit thrushes wintering in a pine plantation landscape. *Journal of Wildlife Management* 66:407–416.

Bryant, D. M. 1979. Reproductive costs in the house martin (*Delichon urbica*). *Journal of Animal Ecology* 48:655–675.

Carbonell, R. and J. L. Tellería. 1999. Feather traits and ptilochronology as indicators of stress in Iberian Blackcaps *Sylvia atricapilla*. *Bird Study* 46:243–246.

Carlson, A. 1998. Territory quality and feather growth in the White-backed Woodpecker *Dendrocopos leucotos*. *Journal of Avian Biology* 29:205–207.

Carrascal, L. M., J. C. Senar, I. Mozetich, F. Uribe, and J. Domènech. 1998. Interactions among environmental stress, body condition, nutritional status, and dominance in Great Tits. *Auk* 115:727–738.

Chen, D. and T. H. Goldsmith. 1986. Four spectral classes of cone in the retinas of birds. *Journal of Comparative Physiology* A159:473–479.

Chen, D., J. S. Collins, and T. H. Goldsmith. 1984. The ultraviolet receptor of bird retinas. *Science* 225:337–340.

Cimprich, D. A. and T. C. Grubb, Jr. 1994. Consequences for Carolina chickadees of foraging with tufted titmice in winter. *Ecology* 75:1615–1625.

Croll, D. A., A. J. Gaston, and D. G. Noble. 1991. Adaptive loss of mass in Thick-billed Murres. *Condor* 93:496–502.

Cuthill, I. C., J. C. Partridge, A. T. D. Bennett, S. C. Church, N. S. Hart, and S. Hunt. 2000. Ultraviolet vision in birds. *Advances in the Study of Behaviour* 29:159–214.

Darwin, C. 1871. *The descent of man and selection in relation to sex*. Murray, London.

De Ridder, E., R. Pinxton, V. Mees, and M. Eens. 2002. Short- and long-term effects of male-like concentrations of testosterone on female European Starlings (*Sturnus vulgaris*). *Auk* 119:487–497.

Del Guidice, G. D., L. D. Mech, and U. S. Seal. 1989. Physiological assessment of deer populations by analysis of urine in snow. *Journal of Wildlife Management* 52:284–291.

Doherty, P. F., Jr. and T. C. Grubb, Jr. 2002. Survivorship of permanent-resident birds in a fragmented forested landscape. *Ecology* 83:844–857.

Doherty, P. F., Jr. and T. C. Grubb, Jr. 2003. Relationship of nutritional condition of permanent-resident woodland birds with woodlot area, supplemental food, and snow cover. *Auk* 120:331–336.

Dolby, A. S. and T. C. Grubb, Jr. 1998. Benefits to satellite members in mixed-species foraging groups: an experimental analysis. *Animal Behaviour* 56:501–509.

Doucet, S. M. 2002. Structural plumage coloration, male body size, and condition in the Blue-black Grassquit. *Condor* 104:30–38.

Doucet, S. M. and R. Montgomerie. 2003. Multiple sexual ornaments in satin bowerbirds: ultraviolet plumage and bowers signal different aspects of male quality. *Behavioral Ecology* 14:503–509.

Eeva, T. and E. Lehikoinen. 1995. Egg shell quality, clutch size and hatching success of the great tit (*Parus major*) and pied flycatcher (*Ficedula hypoleuca*) in an air pollution gradient. *Oecologia* 102:312–323.

Eeva, T., E. Lehikoinen, and M. Rönkä. 1998. Air pollution fades the plumage of the great tit. *Functional Ecology* 12:607–612.

Ekman, J. 1984. Density-dependent seasonal mortality and population fluctuations of the temperate-zone willow tit (*Parus montanus*). *Journal of Animal Ecology* 53:119–134.

Ekman, J. and B. Rosander. 1992. Survival enhancement through food sharing: a means of parental control of natal dispersal. *Theoretical Population Biology* 42:117–129.

Ekman, J., A. Bylin, and H. Tegelstrom. 2000. Parental nepotism enhances survival of retained offspring in the Siberian jay. *Behavioral Ecology* 11:416–420.

Ekman, J., B. Sklepkovych, and B. Tegelstrom. 1994. Offspring retention in the Siberian jay (*Perisoreus infaustus*): the prolonged brood care hypothesis. *Behavioral Ecology* 5:245–253.

Ekman, J., A. Brodin, A. Bylin, and B. Sklepkovych. 1996. Selfish long-term benefits of hoarding in the Siberian jay. *Behavioral Ecology* 7:140–144.

Evans, L. J., E. Ogden, and B. J. M. Stutchbury. 1996. Constraints on double brooding in a Neotropical migrant, the Hooded Warbler. *Condor* 98:736–744.

Fisher, H. I. 1976. Some dynamics of a breeding colony of Laysan albatrosses. *Wilson Bulletin* 88:121–142.

Fitzpatrick, J. W., G. E. Woolfenden, and K. J. McGowan. 1988. Sources of variance in lifetime fitness of Florida scrub-jays. *Proceedings of the International Ornithological Congress* 19:876–891.

Fox, D. L. 1976. Animal biochromes and structural colors. University of California Press, Berkeley, CA.

Frank, R. A. 1990. Sex allocation theory for birds and mammals. *Annual Review of Ecology and Systematics* 21:13–55.

Franzmann, A. W. 1985. Assessment of nutritional status. In R. J. Hudson and R. G. White (eds) *Bioenergetics of wild herbivores*, pp. 239–260. CRC Press, Boca Raton, FL.

Freed, L. A. 1981. Loss of mass in breeding wrens: stress or adaptation? *Ecology* 62:1179–1186.

Fretwell, S. D. 1972. *Populations in a seasonal environment*. Princeton University Press, Princeton, NJ.

Fretwell, S. D. and H. L. Lucas. 1970. On territorial behaviour and other factors influencing habitat distribution in birds. I. Theoretical development. *Acta Biotheoretica* 19:16–36.

Fritts, H. C. 1976. *Tree rings and climate*. Academic Press, New York.

Gaunt, A. S., R. S. Hikida, J. R., Jehl, Jr., and L. Fenbert. 1990. Rapid atrophy and hypertrophy of an avian flight muscle. *Auk* 107:649–659.

Gosler, A. G. 1993. *The great tit*. Hamlyn, London.

Griffiths, R., M. C. Double, K. Orr, and R. J. G. Dawson. 1998. A DNA test to sex most birds. *Molecular Ecology* 7:1071–1075.

Grubb, T. C., Jr. 1989. Ptilochronology: feather growth bars as indicators of nutritional status. *Auk* 106:314–320.

Grubb, T. C., Jr. 1991. A deficient diet narrows growth bars on induced feathers. *Auk* 108:725–727.

Grubb, T. C., Jr. 1992. Ptilochronology: a consideration of some empirical results and "assumptions." *Auk* 109:673–676.

Grubb, T. C., Jr. 1995. On induced anabolism, induced caching, and induced construction as unambiguous indices of nutritional condition. *Proceedings of the Western Foundation of Vertebrate Zoology* 6:258–263.

Grubb, T. C., Jr. and D. A. Cimprich. 1990. Supplementary food improves the nutritional condition of wintering woodland birds: evidence from ptilochronology. *Ornis Scandinavica* 21:277–281.

Grubb, T. C., Jr. and R. Yosef. 1994. Habitat-specific nutritional condition in Loggerhead Shrikes (*Lanius ludovicianus*): evidence from ptilochronology. *Auk* 111:756–759.

Grubb, T. C., Jr. and V. V. Pravosudov. 1994*a*. Ptilochronology: follicle history fails to influence feather growth of an induced feather. *Condor* 96:214–217.

Grubb, T. C., Jr. and V. V. Pravosudov. 1994*b*. Tufted titmouse (*Parus bicolor*). In A. Poole and F. Gill (eds) *The birds of North America*, no. 86. Philadelphia Academy of Sciences, Philadelphia, PA and American Ornithologists' Union, Washington, DC.

Grubb, T. C., Jr., G. E. Woolfenden, and John W. Fitzpatrick. 1998. Factors affecting nutritional condition of fledgling Florida Scrub-jays: a ptilochronology approach. *Condor* 100:753–756.

Grubb, T. C., Jr., T. A. Waite, and A. J. Wiseman. 1991. Ptilochronology: induced feather growth in Northern Cardinals varies with age, sex, ambient temperature, and day length. *Wilson Bulletin* 103:435–445.

Hamilton, W. D. 1964. The genetical evolution of social behaviour. *Journal of Theoretical Biology* 7:1–52.

Hands, H. M., R. D. Brobney, and M. R. Ryan. 1989. Status of the Loggerhead Shrike in the north central United States. Unpublished report to the U.S. Fish and Wildlife Service, Missouri Cooperative Research Unit, University of Missouri, Columbia, MU.

Harder, J. D. 2005. Reproduction and hormones. In C. E. Braun (ed) *Techniques for wildlife investigations and management,* pp. 561–615. The Wildlife Society, Bethesda, MD.

Hay, J. M., P. R. Evans, R. M. Ward, and K. C. Hamer. 2004. Poor nutritional condition as a consequence of high dominance status in the Coal Tit *Parus ater. Ibis* 146:103–107.

Hayes, W. J., Jr. and E. R. Laws. (eds). 1991. *Handbook of pesticide toxicology. Vol. 3, Classes of pesticides*. Academic Press, New York.

Hill, G. E. 1990. Female house finches prefer colorful males: sexual selection for a condition-dependent trait. *Animal Behaviour* 40:563–572.

Hill, G. E. 1991. Plumage coloration is a sexually selected indicator of male quality. *Nature* 350:337–339.

Hill, G. E. and R. Montgomerie. 1994. Plumage colour signals nutritional condition in the house finch. *Proceedings of the Royal Society of London* B 258:47–52.

Hogstad, O. 1990. Dispersal date and settlement of juvenile Willow Tits *Parus montanus* in winter flocks. *Fauna norvegicus Serial C, Cinclus* 13:49–55.

Hogstad, O. 1992. Mate protection in alpha pairs of wintering willow tits, *Parus montanus. Animal Behaviour* 43:323–328.

Hogstad, O. 2003. Strained energy budget of winter floaters in the Willow Tit as indicated by ptilochronology. *Ibis* 145 (online): E19–E23.

Hutchinson, A. E. and R. B. Owen. 1984. Bone marrow fat in waterfowl. *Journal of Wildlife Management* 48:585–591.

Jawor, J. M. and R. Breitwisch. 2004. Multiple ornaments in male northern cardinals, *Cardinalis cardinalis,* as indicators of condition. *Ethology* 110:113–126.

Jawor, J. M., N. Gray, S. M. Beall, and R. Breitwisch. 2004. Multiple ornaments correlate with aspects of condition and behaviour in female northern cardinals. *Animal Behaviour* 67:875–882.

Jenkins, K. D., D. M. Hawley, C. S. Farabaugh, and D. A. Cristol. 2001. Ptilochronology reveals differences in condition of captive White-throated Sparrows. *Condor* 103:579–586.

Jenni, L. and R. Winkler. 1994. *Moult and ageing of European passerines*. Academic Press, London.

Johnson, D. H., T. L. Schaffer, and P. J. Gould. 1992. Incidental catch of marine birds in North Pacific high seas driftnet fisheries in 1990. In International North Pacific Fish Commission (eds), *Symposium on biology, distribution, and stock assessment of species caught in high seas driftnet fisheries in North Pacific Ocean*, November 4–6, 1991. Tokyo, Japan.

Kern, M. D. and R. J. Cowie. 2002. Ptilochronology proves unreliable in studies of nestling Pied Flycatchers. *Ibis* 144:23–29.

Keyser, A. J. and G. E. Hill. 1999. Condition-dependent variation in the blue-ultraviolet coloration of a structurally based plumage ornament. *Proceedings of the Royal Society of London* B 266:771–777.

Keyser, A. J. and G. E. Hill. 2000. Structurally based plumage coloration is an honest signal of quality in male blue grosbeaks. *Behavioral Ecology* 11:202–209.

King, J. R. and M. E. Murphy. 1984. Fault bars in the feathers of White-crowned Sparrows: dietary deficiency or stress of captivity and handling? *Auk* 101:168–169.

Kodric-Brown, A. and J. H. Brown.1984. Truth in advertising: the kinds of traits favored by sexual selection. *American Naturalist* 124:309–323.

Langston, N. E. and S. Rohwer. 1996. Molt-breeding tradeoffs in albatrosses: life history implications for big birds. *Oikos* 76:498–510.

Leary, R. L., F. W. Allendorf. 1989. Fluctuating asymmetry as an indicator of stress: implications for conservation biology. *Trends in Ecology and Evolution* 4:214–217.

Lima, S. L. 1986. Predation risk and unpredictable feeding conditions: determinants of body mass in birds. *Ecology* 67:377–385.

McCarty, J. P. and D. W. Winkler. 1999. Relative importance of environmental variables in determining the growth of nestling Tree Swallows *Tachycineta bicolor. Ibis* 141:286–296.

McGowan, K. J. 1987. Social development in young Florida scrub-jays (*Aphelocoma c. coerulescens*). PhD dissertation. University of South Florida, Tampa, FL.

McGowan, K. J. and G. E. Woolfenden. 1990. Contributions to fledgling feeding in the Florida scrub-jay. *Journal of Animal Ecology* 59:691–707.

MacInnes, C. D., R. A. Davis, R. N. Jones, C. B. Lieff, and A. J. Pakulak. 1974. Reproductive efficiency of McConnel River small Canada Geese. *Journal of Wildlife Management* 38:686–707.

Masello, J. F. and P. Quillfeldt. 2003. Body size, body condition and ornamental feathers of Burrowing Parrots: variation between years and sexes, assortative mating and influences on breeding success. *Emu* 103:149–161.

Mauck, R. A. and T. C. Grubb, Jr. 1995. Petrel parents shunt all experimentally increased reproductive costs to their offspring. *Animal Behaviour* 49:999–1008.

Michener, H. and J. R. Michener. 1938. Bars in flight feathers. *Condor* 40:149–160.

Møller, A. P. 1992. Females prefer large and symmetrical ornaments. *Nature* 357:238–240.

Møller, A. P. 1996. Development of fluctuating asymmetry in tail feathers of the barn swallow *Hirundo rustica*. *Journal of Evolutionary Biology* 9:677–694.

Montiero, L. R. and R. W. Furness. 1996. Molt of Cory's Shearwater during the breeding season. *Condor* 98:216–221.

Mumme, R. L. 1992. Do helpers increase reproductive success: an experimental analysis in the Florida scrub-jay. *Behavioral Ecology and Sociobiology* 31: 319–328.

Murphy, M. E. 1992. Ptilochronology: accuracy and reliability of the technique. *Auk* 109:676–680.

Murphy, M. E. and J. R. King. 1991. Ptilochronology: a critical evaluation of assumptions and utility. *Auk* 108:695–704.

Nilsson, J.-Å. 1989. Establishment of juvenile marsh tits in winter flocks: an experimental study. *Animal Behaviour* 38:586–595.

Nilsson, J.-Å. 1994. Energetic stress and the degree of fluctuating asymmetry—implications for a long-lasting, honest signal. *Evolutionary Ecology* 8:248–255.

Nilsson, J.-Å., H. Källander, and O. Persson. 1993. A prudent hoarder: effects of long-term hoarding in the European nuthatch, *Sitta europaea*. *Behavioral Ecology* 4:369–373.

Nisbet, I. C. T., J. M. Arnold, H. Galbraith, and J. J. Hatch. 2004. Responses of known-aged Common Terns to experimental shortening of the wings. *Waterbirds* 27:13–20.

Nisbet, I. C. T., and E. Cam. 2002. Test for age-specificity in survival of the common tern. *Journal of Applied Statistics* 29:65–83.

Norberg, R. A. 1981. Temporary weight decrease in breeding birds may result in more fledged young. *American Naturalist* 118:838–850.

Norris, K. J. 1990a. Female choice and the evolution of the conspicuous plumage coloration of monogamous male great tits. *Behavioral Ecology and Sociobiology* 26:129–138.

Norris, K. J. 1990b. Female choice and the quality of parental care in the great tit *Parus major. Behavioral Ecology and Sociobiology* 27:275–281.

Nowicki, S., D. Hasselquist, S. Bensch, and S. Peters. 2000. Nestling growth and song repertoire size in great reed warblers: evidence for song learning as an indicator mechanism in mate choice. *Proceedings of the Royal Society of London* B 267:2419–2424.

Nur, N. 1984. The consequences of brood size for breeding blue tits. I. Adult survival, weight change and the cost of reproduction. *Journal of Animal Ecology* 53:479–496.

Ogden, L. J., and B. J. M. Stutchbury. 1996. Constraints on double brooding in a Neotropical migrant, the hooded warbler. *Condor* 98:736–744.

Olson, V. A., and I. P. F. Owens. 1998. Costly sexual signals: are carotenoids rare, risky or required? *Trends in Ecology and Evolution* 13:510–514.

Owen, M. and W. A. Cook. 1977. Variations in body weight, wing length and condition of Mallard *Anas platyrhynchos platyrhynchos* and their relationship to environmental changes. *Journal of Zoology* (London) 183:377–395.

Parsons, P. A. 1990. Fluctuating asymmetry: an epigenetic measure of stress. *Biological Reviews* 65:131–145.

Pennycuick, C. J. 1989. *Bird flight performance: a practical calculation manual.* Oxford University Press, Oxford.

Peters, W. D. and T. C. Grubb, Jr. 1983. An experimental analysis of sex-specific foraging in the downy woodpecker, *Picoides pubescens*. *Ecology* 64:1437–1443.

Pielou, W. P. 1957. A life-history study of the tufted titmouse, *Parus bicolor* Linnaeus. PhD dissertation. Michigan State University, East Lansing, MI.

Polo, V. and L. M. Carrascal. 1999. Ptilochronology and fluctuating asymmetry in tail and wing feathers in Coal Tits *Parus ater. Ardeola* 46:195–204.

Pravosudov, V. V. and T. C. Grubb, Jr. 1999. Effects of inter- and intra-specific dominance on vigilance in avian social groups. *Auk* 116:241–246.

Pravosudova, E. V. and T. C. Grubb, Jr. 2000. An experimental test of the prolonged brood care model in the tufted titmouse (*Baeolophus bicolor*). *Behavioral Ecology* 11:309–314.

Pravosudova, E. V., T. C. Grubb, Jr., and P. G. Parker. 2001. The influence of kinship on nutritional condition and aggression levels in winter social groups of Tufted Titmice. *Condor* 103:821–828.

Reinertsen, R. E. and S. Haftorn. 1986. Different metabolic strategies of northern birds for nocturnal survival. *Journal of Comparative Physiology* 156:655–663.

Rice, D. W. and K. W. Kenyon. 1962. Breeding distribution, history, and populations of North Pacific albatrosses. *Auk* 79:365–386.

Riddle, O. 1908. The genesis of fault bars in feathers and the cause of alteration of light and dark fundamental bars. *Biological Bulletin* 14:328–370.

Riney, T. 1955. Evaluating condition of free ranging red deer (*Cervus elaphus*), with special reference to New Zealand. *New Zealand Journal of Science and Technology* 36B:429–463.

Roff, D. A. 1992. *The evolution of life histories*. Chapman and Hall, New York.

Rogers, C. M. 1987. Predation risk and fasting capacity: do wintering birds maintain optimal body mass? *Ecology* 68:1051–1061.

Rogers, C. M. and J. N. M. Smith. 1993. Life-history theory in the non-breeding period: trade-offs in avian fat reserves. *Ecology* 74:419–426.

Rudolph, D. C., R. N. Conner, and R. R. Shaeffer. 2002. Red-cockaded woodpecker foraging behavior in relation to midstory vegetation. *Wilson Bulletin* 114:235–242.

Rybcynski, R. 1977. Dynamic aspects of bird flocking: the influence of weather and patterns of spatial utilization. PhD dissertation, Cornell University, Ithaca, NY.

Schaefer, R. R. 1996. Red-cockaded woodpecker reproduction and provisioning of nestlings in relation to habitat. MS Thesis. Stephen F. Austin State University, Nacogdoches, TX.

Schaefer, R. R., D. C. Rudolph, R. N. Conner, and D. Daenz. 2005. Red-cockaded woodpecker nutritional status in relation to habitat: evidence from ptilochronology and body mass. *Proceedings of the Fourth Symposium on the red-cockaded woodpecker*. Hancock House, Blaine, Washington DC, in press.

Senar, J. C., J. Figuerola, and J. Domènech. 2003. Plumage coloration and nutritional condition in the great tit *Parus major*: the roles of carotenoids and melanins differ. *Naturwissenschaften* 90:234–237.

Senar, J. C., F. Uribe, J. Domènech, and L. M. Carrascal. 1998. Successive removal of individual rectrices does not influence the subsequent growth of the replacement feathers in the Great Tit *Parus major. Ibis* 140:522–530.

Shawkey, M. D., M. L. Beck, and G. E. Hill. 2003. Use of a gel documentation system to measure feather growth bars. *Journal of Field Ornithology* 74:125–128.

Sklepkovych, B. 1997. The influence of kinship on foraging competition in Siberian jays. *Behavioral Ecology and Sociobiology* 40:287–296.

Slagsvold, T., and J. T. Lifjeld. 1985. Variation in plumage colour of the great tit *Parus major* in relation to habitat, season and food. *Journal of Zoology* 206:321–328.

Stacey, P. B. and J. D. Ligon. 1987. Territory quality and dispersal options in the acorn woodpecker, and a challenge to the habitat-saturation model of cooperative breeding. *American Naturalist* 130:654–676.

Stacey, P. B. and W. D. Koenig. 1990. *Cooperative breeding in birds*. Cambridge University Press, Cambridge.

Stouffer, P. C. and R. O. Bierregaard. 1995. Use of Amazonian forest fragments by understory insectivorous birds. *Ecology* 76:2429–2445.

Stratford, J. A. and P. C. Stouffer. 2001. Reduced feather growth rates of two common birds inhabiting Central Amazonian forest fragments. *Conservation Biology* 15:721–728.

Strickland, D. 1991. Juvenile dispersal in gray jays—dominant brood member expels siblings from natal territory. *Canadian Journal of Zoology* 69:2935–2945.

Strong, A. M. and T. W. Sherry. 2000. Habitat-specific effects of food abundance on the condition of ovenbirds wintering in Jamaica. *Journal of Animal Ecology* 69:883–895.

Stutchbury, B. J. 1994. Competition for winter territories in a Neotropical migrant: the role of age, sex and color. *Auk* 111:63–69.

Sullivan, K. A. 1984*a*. The advantages of social foraging in downy woodpeckers. *Animal Behaviour* 32:16–22.

Sullivan, K. A. 1984*b*. Information exploitation by downy woodpeckers in mixed-species flocks. *Behaviour* 9:294–311.

Takaki, Y., K. Eguchi, and H. Nagata. 2001. The growth bars on tail feathers in the male Styan's Grasshopper Warbler may indicate quality. *Journal of Avian Biology* 32:319–325.

Talloen, W., L. Lens, and S. Van Dongen. 2005. Avian feather development under environmental stress: lead pollution effects on growth patterns and realized phenotypes. *Ibis*, in press.

Tarboton, W. R. 1981. Cooperative breeding and group territoriality in the Black Tit. *Ostrich* 52:216–225.

Trivers, R. L. 1972. Parental investment and sexual selection. In B. Campbell (ed.), *Sexual selection and the descent of man 1871–1971*, pp. 136–179 Aldine, Chicago, IL.

Trivers, R. L., and D. E. Willard. 1973. Natural selection and paternal ability to vary the sex ratio of offspring. *Science* 179:90–92.

van Ort, H. and K. A. Otter. 2005. Natal nutrition and the habitat distributions of male and female black-capped chickadees. *Canadian Journal of Zoology* 83: in press.

Veit, A. C. and I. L. Jones. 2004. Timing and patterns of growth of red-tailed tropicbird *Phaethon rubricauda* tail streamer ornaments. *Ibis* 146: 355–359.

Velando, A. 2002. Experimental manipulation of maternal effort produces differential effects in sons and daughters: implications for adaptive sex ratios in the blue-footed booby. *Behavioral Ecology* 13:443–449.

Von Schantz, T., S. Bensch, M. Graham, D. Hasselquist, and H. Wittzell. 1999. Good genes, oxidative stress and condition-dependent sexual signals. *Proceedings of the Royal Society of London B* 266:1–12.

Waite, T. A. 1990. Effects of caching supplemental food on induced feather regeneration in wintering Gray Jays *Perisoreus canadensis*: a ptilochronology study. *Ornis Scandinavica* 21:122–128.

Waite, T. A. and D. Strickland. 1997. Cooperative breeding in gray jays: philopatric offspring provision juvenile siblings. *Condor* 99:523–525.

Warren, R. J., R. L. Kirkpatrick, A. Oelschlaeger, K. E. Webb, Jr., and J. B. Whelan. 1982. Energy, protein, and seasonal influence on white-tailed deer fawn nutritional indices. *Journal of Wildlife Management* 46:302–312.

Weise, C. M. and J. R. Meyer. 1979. Juvenile dispersal and development of site-fidelity in the black-capped chickadee. *Auk* 96:40–55.

White, D. W. and E. D. Kennedy. 1992. Growth of induced feathers in photostimulated American Tree Sparrows. *Condor* 94:543–545.

White, D. W., E. D. Kennedy, and P. C. Stouffer. 1991. Feather regrowth in female European Starlings rearing broods of different sizes. *Auk* 108:889–895.

Wolf, B. O. and G. E. Walsberg. 1996. Thermal effects of radiation and wind on a small bird and implications for microsite selection. *Ecology* 77:2228–2236.

Wood, H. B. 1950. Growth bars in feathers. *Auk* 67:486–491.

Woolfenden, G. E. and J. W. Fitzpatrick. 1990. Florida scrub-jays: a synopsis after 18 years of study. In P. B. Stacey and W. D. Koenig (eds), *Cooperative breeding in birds*, pp. 241–266. Cambridge University Press, Cambridge.

Woolfenden, G. E. and J. W. Fitzpatrick. 1996. Florida Scrub-jay (*Aphelocoma coerulescens*). In A. Poole and F. Gill (eds), *The birds of North America*, No. 228. Academy of Natural Sciences, Philadelphia, PA and American Ornithologists' Union, Washington, DC.

Yosef, R. 1992. Territoriality, nutritional condition, and conservation in loggerhead shrikes (*Lanius ludovicianus*). PhD dissertation. The Ohio State University, Columbus, OH.

Yosef, R. 1997. On habitat-specific nutritional condition in Graceful Warblers *Prinia gracilis*: evidence from ptilochronology. *Journal für Ornithologie* 138:309–313.

Yosef, R. and T. C. Grubb, Jr. 1992. Territory size influences nutritional condition in nonbreeding loggerhead shrikes (*Lanius ludovicianus*): a ptilochronology approach. *Conservation Biology* 6:447–449.

Yosef, R. and T. C. Grubb, Jr. 1995. Resource dependence and territory size in Loggerhead Shrikes (*Lanius ludovicianus*). *Auk* 111:465–469.

Zahavi, A. 1975. Mate selection—a selection for a handicap. *Journal of Theoretical Biology* 53:205–214.

Zuberbier, G. M. and T. C. Grubb, Jr. 1992. Ptilochronology: wind and cold temperatures fail to slow induced feather growth in captive White-breasted Nuthatches *Sitta carolinensis* maintained on an ad libitum diet. *Ornis Scandinavica* 23:139–142.

Index